Approaches to Improve Agricultural Water Productivity

Approaches to Improve Agricultural Water Productivity

A.K. Singh
Principal Scientist (Horticulture) and Head

S.S. Mali
Scientist, Soil and water Conservation Engineering

B. Das
Senior Scientist, Fruit Science

P. Bhavana
Scientist, Plant Breeding

P.R. Kumar
Senior Scientist, Plant Breeding

B. Sarkar
Senior Scientist, Farm Machinery

B.P. Bhatt
Director

ICAR-RCER, Research Centre Ranchi
Jharkhand, India

SSPH

SATISH SERIAL PUBLISHING HOUSE
403, Express Tower, Commercial Complex
Azadpur, Delhi-110033 (India)
Phone : 011-27672852, Fax : 91-11-27672046
e-mail : info@satishserial.com, hkjain1975@yahoo.com
Website : www.satishserial.com

Published by :

SATISH SERIAL PUBLISHING HOUSE

403, Express Tower, Commercial Complex, Azadpur, Delhi-110033 (INDIA)
Phone : 011-27672852 Fax : 91-11-27672046
E-mail : info@satishserial.com, hkjain1975@yahoo.com

ISBN 978-93-85055-80-5

Composed, Designed & Printed in India

Preface

Increasing number of regions are facing severe water shortage. It is estimated that about one-third of the world's population lives in water scarce or extremely water scarce areas and by 2025 it is expected to grow to two-thirds. Agriculture is the largest water consumer worldwide and fulfilling agricultural water demand will soon become a major challenge in most parts of the world. The ideal response to this would be implementation of improved technologies, bringing in changes in water use practices and water resource management. Development and adoption of water-efficient technologies can improve the situation in respect of excessive water use in agriculture. Several technological innovations aiming at "more crop for every drop" are proposed for increasing water use efficiency which could also lead to increase in water productivity in the agricultural sector.

Growing water scarcity has led to a renewed interest in water productivity as a potentially useful concept to identify where improvements in agricultural production can be made. The challenge is to improve land and water productivity, evolve effective management alternatives to cope up with erratic climatic conditions and to provide adequate support and infrastructural facilities. Variety of technological options have been developed in the fields of water management, integrated farming systems, soil health management, crop protection and agroforestry. These options not only can improve the economic water productivity substantially but also cater to the needs of small, marginal and resource poor farmers of the country. As a consequence, a lot of hope has been bestowed upon Eastern Plateau and Hill Region in the scheme of things which would usher the country into the second phase of green revolution.

This publication brings out new approaches, methods, concepts and findings to address the growing and important issue of improving water productivity. In the context of growing water scarcity and low water productivity, this publication would be highly useful to scientists, policy makers and farmers.

Authors

About the Authors

Dr. A K Singh, is a Principal Scientist and Head at ICAR-RCER, Research Centre, Ranchi. He has 24 years of research experience in horticultural crops and has developed 35 varieties/F1 hybrids of different vegetables for different parts of the country. He also contributed to standardization of package of practices for important vegetable crops cultivated in hill and plateau region of North Eastern India. He has published 66 research papers in different national and international journals and has authored one book, four book chapters and 47 technical bulletins/ folders. He is recipient of number of prestigious awards like Crop Research Award, Senior Scientists Gold Medal Award-2009 by Madhawi Shyam Educational Trust and International Consortium of Contemporary Biologists, Ranchi. He is recipient of the prestigious Fakhruddin Ali Ahmed Award of the ICAR for his outstanding contribution in improving the livelihood of tribal farmers of the region.

Dr. Santosh S. Mali, received his MSc and PhD from Division of Agricultural Engineering, IARI, New Delhi. Presently he is working as Scientist (SWCE) at the Ranchi centre of ICAR-RCER, Patna. Before Joining ICAR he served as Scientist-B in National Institute of Hydrology, Roorkee. His current research focuses on micro-irrigation, climate change impact assessment and surface water hydrology with special emphasis on application of remote sensing and GIS in natural resource management. His areas of interest includes water footprints of agricultural commodities and groundwater flow modelling. He has successfully completed one and is part of many externally funded projects. He is also involved in ACIAR-IWMI supported international collaborative project on improving dry season irrigation in Gangetic plains. Dr Mali is a life member of Indian Society of Agricultural Engineers and Indian Association of Soil and Water Conservationists.

Dr. Bikash Das, is presently working as Senior Scientist (Horticulture) at Ranchi centre of the ICAR-RCER, Patna. He has 12 years of research experience in the field of development of fruit based production systems. His work emphasized on efficient utilization of natural resources in the tribal dominated region of Jharkhand and West Bengal. He is known for providing the technological backstopping to different stakeholders working on horticultural development in these states. He has published 41 research papers in national and international journals, authored a book, seven book chapters and six technical bulletins. He is receipient of young scientist gold medal award by Madhawi Shyam Educational Trust and International Consortium of Contemporary Biologists, Ranchi. He has also received the prestigious Fakhruddin Ali Ahmed Award of the ICAR for his outstanding contribution in improving the livelihood of tribal farmers of the region.

Dr. (Mrs) P. Bhavana, is presently a Scientist (Plant Breeding) at ICAR-Research Complex for Eastern Region, Research Centre, Ranchi, Jharkhand, India; obtained her Masters from Acharya N.G. Ranga Agricultural University, Hyderabad and Ph.D. (Genetics) from Indian Agricultural Research Institute, New Delhi with CSIR-UGC JRF in Life Sciences. Her research interest is bacterial wilt resistance breeding in brinjal and tomato. She has the credit in the release of two varieties and one hybrid of tomato, two varieties of chilli, one variety each of capsicum, bottle gourd, bitter gourd and ridge gourd through State Variety Release Committee, Govt. of Jharkhand. Dr Bhavana is a life member of The Horticultural Society of India and Indian Society of Vegetable Sciences.

Dr. Priya Ranjan Kumar, is presently working at ICAR-Research Complex for Eastern Region, Research Complex, Research Centre, Ranchi (Jharkhand) as a Senior Scientist (Plant Breeding). He did his Ph.D. in Seed Science and Technology from Indian Agricultural Research Institute, New Delhi. He has seventeen years of experience in the field of seed production and quality assurance of a wide gamut of crops including field crops, vegetables, cash crops (like tobacco) and fruits (papaya). As a plant breeder he has developed a hybrid of bell pepper, a CMS line of bell pepper and has five varieties of wheat to his credit. Dr Kumar has published 20 research papers and contributed chapters to several books.

Dr. Bikash Sarkar, is working as Senior Scientist (Farm Machinery and Power Engineering) at ICAR Research Complex for Eastern Region, Patna. He did his graduation in Mechanical Engineering. He received Post Graduation (M.S.) with specialization in Manufacturing Management from BITS, Pilani. He received Ph.D. degree from the Centre for Energy Studies, Indian

Institute of Technology, Delhi. He has published over 40 research papers in Journal of repute. His areas of research interest are Farm Mechanization and protected cultivation technology.

Dr. B P Bhatt, is the Director of the ICAR-Research Complex for Eastern Region, Patna. He started his career as lecturer (Forestry) at H.N.B Garhwal University, Srinagar and later served as Joint Director and Principal Scientist at ICAR-Reserch Complex for North Eastern Hill Region, Barapani. He has 12 years of teaching experience in the field of agroforestry and has supervised seven PhD Students. Dr. Bhatt has research experience in the fields of agroforestry, integrated farming systems and natural resource management. He has more than 200 publications in the form of research papers, book chapters, bulletins, policy documents and training manual to his credit. He is recipient of the prestigious Fakhruddin Ali Ahmed Award of the ICAR for his outstanding contribution on Intensive Integrated Farming Systems for Livelihood Security in NEH region. He is also a national fellow of the National Academy of Agricultural Sciences in the field of Natural Resource Management and Fellow of the National Institute of Ecology.

Editors and Contributors

Editors:

Mali S S	Scientist (Soil and water Conservation Engineering), ICAR-RCER, Research Centre Ranchi, Jharkhand
Das B	Senior Scientist (Fruit Science), ICAR-RCER, Research Centre Ranchi, Jharkhand
Bhavana P	Scientist (Plant Breeding), ICAR-RCER, Research Centre Ranchi, Jharkhand
Kumar P R	Senior Scientist (Plant Breeding), ICAR-RCER, Research Centre Ranchi, Jharkhand
Sarkar B	Senior Scientist (Farm Machinery),ICAR-RCER, Patna, Bihar
Singh A K	Principal Scientist & Head, ICAR-RCER, Research Centre Ranchi, Jharkhand
Bhatt B P	Director, ICAR-RCER, Patna, Bihar

Contributors:

Chapter 1 Basic Concepts of Water Productivity

Mali S S, Scientist, ICAR-RCER, Research Centre Ranchi, Jharkhand

Chapter 2 Climate Change, Water and Agriculture

Islam A, Principal Scientist, NRM Division, ICAR, New Delhi

Naik S K, Senior Scientist, ICAR-RCER, Research Centre Ranchi, Jharkhand

Chapter 3 Water Harvesting and Management

Mali S S, Scientist, ICAR-RCER, Research Centre Ranchi, Jharkhand

Singh A K, Senior Scientist, Centre for Protected Cultivation Technology, ICAR-IARI, New Delhi

Kumar A, Senior Scientist, ICAR-Research Complex for Eastern Region, Patna, Bihar

Chapter 4 **Irrigation Management in Fruit Crops**

Dhakar M K, Scientist, ICAR-RCER, Research Centre Ranchi, Jharkhand

Mali S S, Scientist, ICAR-RCER, Research Centre Ranchi, Jharkhand

Chapter 5 **Precision Farming and GIS**

Awani K Singh, Senior Scientist, Centre for Protected Cultivation Technology, ICAR-IARI, New Delhi

Manibhushan, Scientist, ICAR-Research Complex for Eastern Region, Patna, Bihar

Chapter 6 **Fruit Based Production Systems**

Das B, Senior Scientist, ICAR-RCER, Research Centre Ranchi, Jharkhand

Dhakar M K, Scientist, ICAR-RCER, Research Centre Ranchi, Jharkhand

Chapter 7 **Vegetable Based Production Systems**

Singh A K, Principal Scientist and Head, Research Centre Ranchi, ICAR-RCER, Patna, Bihar

P Bhavana, Scientist, ICAR-RCER, Research Centre Ranchi, Jharkhand

Das B, Senior Scientist, ICAR-RCER, Research Centre Ranchi, Jharkhand

Chapter 8 **Water Productivity of livestock**

Dey A, Principal Scientist, Division of Livestock & Fishery Management ICAR-Research Complex for Eastern Region, Patna, Bihar

Sharma K, Principal Scientist, ICAR-Research Complex for Eastern Region, Patna, Bihar

Gupta J J, Principal Scientist, ICAR-Research Complex for Eastern Region, Patna, Bihar

Giri S C, Senior Scientist, Regional Centre, ICAR-Central Avian Research Institute, Bhubaneswar, Odisha

Chakrabarti A, Senior Scientist, ICAR-RCER, Research Centre Ranchi, Jharkhand

Chapter 9 **Crop Diversification Options**

Kumar S, Ex Principal Scientist and Head, ICAR-RCER, Research Centre Ranchi, Jharkhand

Kumar Jai, Assistant Professor-cum – Junior Scientist, AICRP on M&APs, Department of Forest Products & Utilization, Birsa Agricultural University, Ranchi, Jharkhand

Sharma J P, Ex Principal Scientist, ICAR-RCER, Research Centre Ranchi, Jharkhand

M. Nedunchezhiyan, Regional Centre of ICAR-Central Tuber Crops Research Institute, Bhubaneswar, Odisha

Srkar P K, Scientist, ICAR-RCER, Research Centre Ranchi, Jharkhand

Sarkar P, Scientist, ICAR-RCER, Research Centre Ranchi, Jharkhand

Sinha A, Scientist, Institute of Forest Productivity, Lalgutwa, Ranchi, Jharkhand

Chapter 10 **Enterprise Diversification Options**

Bhattcharya A, Ex PS and Direcctor, ICAR-Indian Institute of Natural Resins and Gums, Ranchi, Jharkhand

Giri S, Scientist-C, Central Tasar Research & Training Institute, Nagri, Ranchi, Jharkhand

Kumar P R, Senior Scientist, ICAR-RCER, Research Centre Ranchi, Jharkhand

Chapter 11 **Integrated Farming Systems**

Jha B K, Senior Scientist, ICAR-RCER, Research Centre Ranchi, Jharkhand

Kumar S, Principal Scientist, ICAR-Research Complex for Eastern Region, Patna, Bihar

Pan R S, Principal Scientist and Head, ICAR-RCER, Research Centre Ranchi, Jharkhand

Chapter 12 **Conservation Agriculture**

Malik R K, Senior Agronomist, CSISA, CIMMYT, New Delhi

Chapter 13 **Crop Improvement for Water and Heat Stress**

Bhavana P, Scientist, ICAR-RCER, Research Centre Ranchi, Jharkhand

Gutam S, Senior Scientist, ICAR-RCER, Research Centre Ranchi, Jharkhand

Chapter 14 Pest, Disease and Weed Management

Choudhary J S, Scientist, ICAR-RCER, Research Centre Ranchi, Jharkhand

Maurya S, Senior Scientist, ICAR-RCER, Research Centre Ranchi, Jharkhand

Mishra J S, Principal Scientist and Head, Division of Crop Research, ICAR-RCER, Patna, Bihar

Chapter 15 Soil Health Management

Naik S K, Senior Scientist, ICAR-RCER, Research Centre Ranchi, Jharkhand

Chapter 16 Technology Upscaling and Participatory Approach

Singh H S, Head, Central Horticultural Experiment Station, ICAR–IIHR, Bhubaneswar, Odisha

Kumar U, Principal Scientist, Division of Socio-economic and Extension, ICAR-RCER, Patna, Bihar

Contents

Contents

1

Basic Concepts of Water Productivity

Societies have to make the best use of the limited water resources in our world. Water is the vital resource to sustain civilizations and economic development and most importantly, agriculture. Although available in abundance in some regions, increasing number of river basins of arid and semi-arid regions are facing problem of water scarcity as well as deterioration of water quality, due to population growth and increased water use (Vorosmarty *et al.*, 2010). About one-third of the world population is suffering from different levels of water stress, while water scarcity is likely to affect two-thirds of the world population over the next decades (Oki and Kanae, 2006). Rapid industrialization and urbanization, expansion of agriculture and climate change are the major drivers of water scarcity all over the world.

To feed an ever increasing population with more water demanding lifestyles, more water is required in both agriculture as well as domestic sectors. Increased water demand for agriculture will strain terrestrial and aquatic ecosystems and intensify competition for water resources. Improving physical water productivity in agriculture reduces the need for additional water and land in irrigated and rainfed systems and is thus a critical response to increasing water scarcity. Water saving achieved from improved water productivity in agriculture will serve the need to sustain ecosystems. Also, increasing competition from industrial sector is further adding stress on country's water resources.

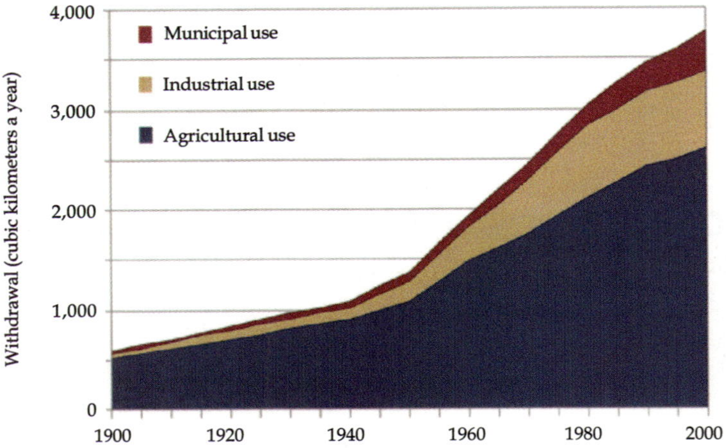

Fig. 1.1 Sector wise global water withdrawal and population trend

Adopting proven agronomic and water management practices could raise the water productivity because raising land productivity generally leads to increases in water productivity. Other than direct benefits accrued in terms of improved yield and water saving there are several indirect benefits of improved water productivity. Water productivity gains are context dependent and can be properly assessed only by taking an integrated basin perspective. It is logical to assess the other benefits in terms of employment and income generation, nutrition enrichment and new opportunities created for women.

Increasing demand for livestock and fish products has raised the demand for allocation of water to these sectors. Reduction in river flows is threatening the existence of some fish species consequently affecting the social and ecological values generated by fisheries sector. There is good scope to improve physical and economic water productivities in livestock and aquaculture sectors. Improvement in water productivity can be achieved through careful selection of feed sources, improving the quality and value of produce through processing and integrating fisheries and livestock into farming systems. Targeting high water productivity can reduce investment costs by reducing the amount of water that has to be withdrawn. Over the past few years, the concept of productivity of water in agriculture has gained ground owing to increasing scarcity of irrigation water from physical and economic perspectives.

Concept of Water Productivity

Water productivity has been given different definitions by different authors, often according to the scale of the plant, plot of land or watershed they were investigating or the purpose of their study. Molden (1997) defined water productivity as the physical mass of production or the economic value of production measured against gross inflow, net inflow, depleted water, process depleted water, or available water. Water productivity is usually estimated as the amount of agricultural output produced per unit of water consumed. Mathematically water productivity is expressed as:

"Water Productivity (kg/m3 or S/m3)" = "Output drived from water use"

In broader sense it is defined as the ratio of the net benefits from crop, forestry, fishery, livestock, and mixed agricultural systems to the amount of water required to produce those benefits. Water productivity reflects the goal of producing more food, income, livelihoods, and ecological benefits at less social and environmental cost per unit of water. Water used in a particular use may be water delivered to a use or depleted (consumed) by a use. Water is considered as consumed is largely site-specific, as some of it may be reutilized in other production processes. Two specific types most widely used in literature are physical water productivity defined as the ratio of the mass of agricultural output to the amount of water used and economic water productivity defined as the value derived per unit of water used.

The denominator of the water productivity equation is expressed as water either supplied or depleted. Water is depleted when it is consumed by evapotranspiration, is incorporated into a product, flows to a location where it cannot be readily reused (to saline groundwater, for example), or becomes heavily polluted (Seckler 1996; Molden and others 2003).

Some Performance Indices

Water productivity is one of the many indices used in the assessment of effectiveness of the production systems as a whole. Indices aid in decision making processes at the managerial level to decide upon the water resource management and allocation strategies. Comparisons of the different production systems within a single unit system is possible with the indices that are derived from the system parameters. Assessing incremental welfare per unit of water and identification of opportunities to improve the net gain can be achieved with the use of these indices. Some indices used in the agricultural production systems are mentioned below :

- Yield per hectare (land productivity) expressed in t/ha or q/ha is the most commonly used indicator to assess the performance of the system.
- Transpiration coefficient: (TC). This coefficient, which is measured as the ratio of the weight of water absorbed to the weight of dry matter produced, was introduced by Briggs and Shantz (1913, 1914) and is expressed as m^3/kg. This index particularly reflects the performance of a plant.
- Rain use efficiency(RUE) is ratio of annual production to the amount of annual rainfall received by the field and has the same unit as that of water productivity (kg/m^3). Since total rainfall received is considered as water applied, the denominator accounts for runoff and groundwater recharge. Rain use efficiency (RUE) has been proposed by Le Houérou (1984).
- Water use efficiency (Viets, 1962) is another most widely used terminology in the field of agricultural production. It is the ratio between dry matter produced and the amount of water evaporated and transpired. It has the same unit as that of RUE (kg/m^3). It describes the relation between production and water loss.

Water productivity analysis can be applied to crops, livestock, tree plantations, fisheries, and mixed systems at selected scales. The objectives of water productivity analysis range from assessing agricultural production (kilograms of grain per unit of water depleted by a crop on a field) to assessing incremental welfare per unit of water used in the agricultural sector. Crop physiologists defined water use efficiency as carbon assimilated and crop yield per unit of transpiration (Viets 1962), and then later as the amount of produce per unit of evapotranspiration. Irrigation specialists have used the term water use efficiency to describe how effectively water is delivered to crops and to indicate the amount of water wasted. But this concept provides only a partial and sometimes misleading view because it does not indicate the benefits produced, and water lost by irrigation is often gained by other uses (Seckler *et al.*, 2003).

Irrigation Efficiency, Water Use Efficiency and Water Productivity

The terminologies, Irrigation efficiency, Water use efficiency and water productivity seem to cause some confusion among researchers. The fact is, all these terms do exist and have different definitions and applications. In some of the previous literatures researchers considered WP and WUE are synonymous. However, there is substantial difference between irrigation efficiency, water use efficiency and the water productivity. Irrigation efficiency aims at assessing the performance of the irrigation system. Molden (1997) mentioned that productivity takes different forms with different units but efficiency has only one form (dimensionless). The most recent paper on that subject (Heydari, 2014) indicated that WP is distinct from WUE as WP refers to crop production in relation to total water consumed while the WUE is a dimensionless ratio of total amount of water used to the total amount of water applied, "as WP terms are not dimensionless, i.e. cannot be categorized in efficiency terms, they are just some ratios with different units in the numerator and denominator".

In literature published during the 1960s, the term 'water productivity' was referred to as 'water use efficiency'. Some of the confusions in the definition of water productivity come from the fact that some researchers use it interchangeably with water use efficiency. The recent literature shows that what previously was wrongly defined as Water Use Efficiency has been renamed as "Water Productivity" in the early 1980's.

The context of application of these terms is completely different. Irrigation is more important to the irrigation engineers to assess the performance of the irrigation systems while the term water use efficiency is more useful to crop physiologist to assess the efficiency with which crop used the applied water and is applicable at individual plant scale. Water productivity is a broader term that encompasses all the benefits from the system as whole in to one representative index. Following table makes the difference more clear.

Table 1.1 Difference between IE, WUE and WP

Irrigation Efficiency	Water Use Efficiency	Water Productivity
Ratio of water consumed by crops to water diverted from the source.	Dimensionless ratio of total amount of water used to the total amount of water applied	Refers to crop production/ benefits in relation to total water consumed
Applies to irrigation system	Applies to crop	Its related to benefits from a system
Used to evaluate the performance of water system	Performance of crop	Performance of production system as whole
Objective-water saving	Just an assessment of amount of water taken up by the plant	Objective – Getting best returns from applied water
Non-dimensional	Non-dimensional	Has dimension of $[M^1L^{-3}T^0]$
Considers losses – seepage, soil evaporation	Does not consider losses	Loss accounting depends on context – Supply or depletion water productivity

Improved IE and WUE contributes in improving - WP

Water productivity as value per unit of water

Prima facie it appears that the term 'water productivity' deals with the physical yields obtained per unit of water. If we shift the context to economic returns, the broader sense of this term comes into picture. Maximizing the net benefits or deriving greater value per unit of water has key implications for farmer decisions, economic growth and the environment. Compared to physical water productivity, there is good scope for increasing value (economic returns) per unit of water used in agriculture. This can be achieved through evolving strategies for increasing yield per unit of supply or depletion. The first step to increase the value of water can be to cultivate high value crops, instead of low value crops which offer less economic returns. Other better alternative is to lower the costs of agricultural inputs, which will increase the net benefit per unit of water used.

Integration of fish and livestock in to traditional single commodity production systems can ensure multiple uses of water and significantly improve the value derived per unit of water used. Allocation of water to one system may come at the cost of reduced water availability for other production system. Therefore, understanding values derived from different combination of enterprises will help to understand where there are win-win situations and what trade-offs will have to be made.

Water productivity and Associated Benefits

Improving water productivity offer wide range of benefits, with quantifiable and non-quantifiable elements. Agricultural output, which can be evaluated for physical, nutritional or monetary benefit, within the bounds of the scale and time period being considered. Other benefits associate with crop production, for example, can be fodder for livestock and organic matter for soil quality improvement. Some cover crops also provide protection form soil erosion. Current approaches for assessing the agricultural water-use benefits generally ignore secondary goods and services.

Apart from direct benefits like water saving and increased yield, it offers several long term outcomes of livelihood improvement and better ecosystem services. Following chart explains the benefits offered by improved water productivity (Fig. 1.2). Some of these benefits can be quantified through technical and social data. Other benefits are inherent, but their quantification is difficult, rather systematic procedure to quantify these is not available. The potential benefits of improving water productivity are summarised as:

- Non-grain benefits of water use in crop production such as the use of crop residues as fodder and/or mulch.
- Benefits from by-products of livestock and fish production and their role as food supplements for livestock and fish production systems or as inputs to enhance soil fertility.
- Benefits from ecosystem goods and services (biodiversity, ecosystem integrity, habitat maintenance) and socio-cultural benefits, such as aesthetics and cultural importance, derived from hydrologic flows in agricultural water use systems.

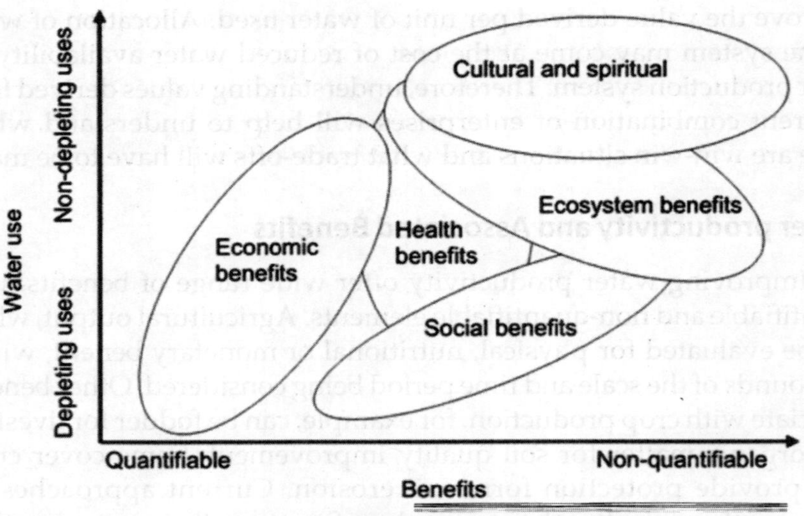

Fig 1.2 Water use and associated benefits (Gichuki *et al.*, 2015)

Water productivity improvement can enable poor and marginalized people to gain access to water and use it more effectively. Increased water productivity may benefit poor people through multiplier effects on food security, employment, and income. The current focus of water productivity has evolved to include the benefits and costs of water used for agriculture in terrestrial and aquatic ecosystems. Water productivity analysis can be seen as part of an ecosystem approach for managing water. Rain, natural flows, withdrawals, and evaporation support terrestrial and aquatic ecosystems, which produce numerous services for people. The primary service of agro ecosystems is food and fibre production, but other important services are produced as well.

Assessment of Water Productivity

Water productivity is expressed as a ratio of two quantities. It has a numerator and a denominator. In case of fodder crops the numerator can be total above ground dry matter, for other crops it can be edible weight of food crops. When we talk about economic water productivity the monetary benefits emanating from the water use becomes the numerator. At ecosystem level we consider the value of other benefits like improved livelihood and services rendered to ecosystem, however, this is very complex to estimate. The denominator of the estimation typically include the amount of water – supplied or depleted. If the amount of water applied (supplied to field) is used as denominator it is termed as irrigation water productivity. The depleted water consists of evapotranspiration and water incorporated into a product and other flows which cannot be readily reused (to saline groundwater, for example), or become heavily polluted. Following schematic explains the parameters used in estimation of water productivity.

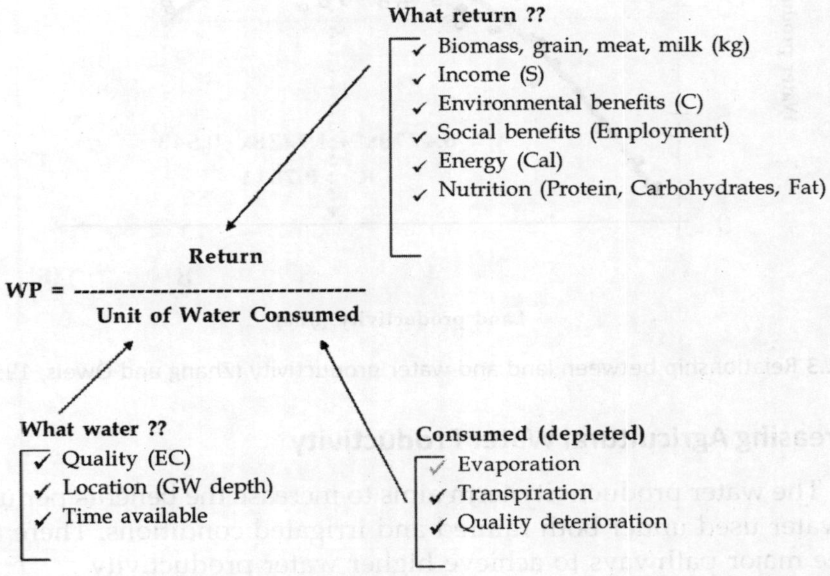

Water productivity assessment framework

Tradeoffs between land and water productivity

Within a certain 'favourable' limit of amount of water applied, the absolute yield per unit of land continues to increase and there is increase in the land productivity with increasing depth of water applied. 'Favourable' because too much water application may result in water logging and reduced crop yields. As the yield increases, water productivity also increases reaching a point where it becomes maximum (Max WP). This point of inflexion is important in the water scarce regions where the focus is to get most out of every drop of water. At this point benefits derived per unit of water are highest is of much importance from water productivity point of view. As we go further, although the yield increases, the water productivity will decrease. This is mainly because the yield increase is not in the direct proportion of the quantity of water applied.

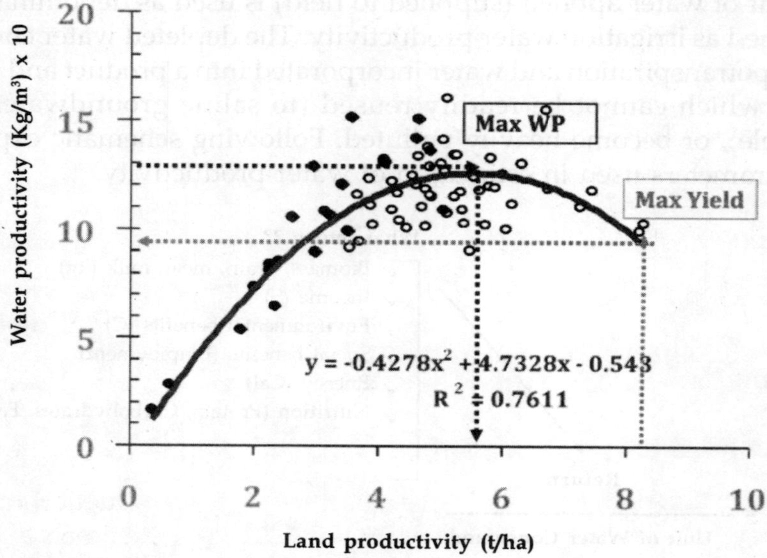

Fig 1.3 Relationship between land and water productivity (Zhang and Oweis, 1999)

Increasing Agricultural Water Productivity

The water productivity term aims to increase the benefits per unit of water used under both rainfed and irrigated conditions. There are three major pathways to achieve higher water productivity :

1. Increasing the marketable yield of the crops
2. Reducing the outflows/loss of water from the system as whole
3. Making effective use of rainfall, soil moisture and marginal quality water.

Technological options that lead to improvement of the on-farm management of soil, water and crop, through the application of the best crop management practices will improve the water productivity. These practices permit to use less water for irrigation, decrease evaporation losses, optimize fertilizer supply, allow better pest control, minimize energy consumption and improve soil conditions.

Genetic enhancement of the crops for yield improvement and natural resource management has made an important contribution to increases in agricultural water productivity. Efforts are required to reduce losses such as evaporation and deep percolation so as to reduce the denominator term in the water productivity equation. Minimum tillage, zero tillage, efficient use of soil moisture and improving nutrient status of soil can improve the biomass productivity of water. Supplemental irrigation, deficit irrigation and irrigation during critical stages of crop growth can significantly influence the crops yield leading to improvement in water productivity. Crop breeding programs for increasing the proportion of economic biomass, reducing the growing period, and tolerance to higher and lower temperatures as well as resistance to range of diseases and pests can substantially gain on water productivity. Economic water productivity gains can be achieved through improvements in soil fertility and management of rainwater to reduce evaporation and divert more flows to transpiration.

Water Productivity of Livestock

Livestock is an integral part of the agricultural production systems. Water is required for animals mainly for the food that animals consume and water required to maintain the hygienic conditions at the barns. Compared to water required for production of animal feed, the drinking water requirements of livestock are negligible. Therefore, by improving the management practices for crops, carefully selecting the kind of feed, efficient reuse of crop residues, and improving the feed processing system will lead to increased water productivity. Livestock improves the productivity of the system as a whole by complementing other production systems. Livestock add value in many ways to production systems and play an important role in livelihood strategies, contributing to overall productivity and welfare gains. Other benefit include support to the smallholders in both irrigated and rainfed systems. Processing the animal products and value addition can improve the water productivity of livestock many folds.

Water productivity of Aquaculture

Although consumption of water by fishes and other aquaculture species is negligible, there are two major components of water use; water required for feed and the blue water required for aquaculture. In case of aquaculture the water productivity is the mass or value of the aquaculture produce divided by the amount of water. Water use includes the amount of water required in feed production and water consumed by the fish. Evaporation from the pond is also included in water use as this can be a major component in arid and semi-arid regions. Irrigation reservoirs can be used for culturing fish which will give an additional income to the farmer. Adoption of cage aquaculture in irrigation canals is also a potential option, given the operational issues. The values and livelihood benefits of fisheries are high and often ignored or underestimated, but considering only the values of fish produce would grossly underestimate the value of water in these aquatic ecosystems. The water productivity of fisheries systems needs to be considered in terms of the ecosystem services and livelihoods supported per unit of water. Thus maintenance of wetlands and biodiversity should be considered potential benefits of leaving water in these aquatic ecosystems (Molden and Oweis 2015).

Agricultural water management practices can provide multiple ecosystem services beyond food production. Practices that reduce environmental costs and enhance ecosystem services increase the value derived from agricultural water management (Matsuno *et al.*, 2006).

Incentives and Policy

Many innovative and practicable technological options have been developed to improve water productivity but adoption rates remain low. There are other major issues like water productivity itself is yet to feature prominently among agricultural producers. Farmers rarely manage to increase water productivity; rather they manage to make their entire enterprise profitable. Access to low-cost supplies of sufficient water enable high levels of productivity and reduce risk. Low adoption of these technologies is mainly limited by the economic status of the farmer to pay for a management practice or technology and there is concern in farmers mind over the payback on the investment. Adoption of a technology is also governed by the risk associated with the technology as the returns from a particular strategy vary greatly in time and depends on market and climatic aberrations. Lack of technical know-how of a technology could be another important reason that discourages the uptake of potential technology.

The incentives for improving the water productivity can be categorised in to direct and indirect incentives. The indirect incentives comes from the farmers themselves which is governed by the instinct of the farmer to produce more or to get more income while direct incentives are from governments and other funding agencies. Tendency to retain the water supply for more production and its use in forthcoming season is a type of indirect incentive. One strategy is to compensate the farmers for investing in water saving and profitability enhancing farming technologies. Many direct incentives will come from outside the water sector. These include address issues of the vulnerability and risk and markets and profitability of the agriculture enterprise and the equity and welfare of stakeholders.

□□□

2

Climate Change, Water and Agriculture

Agriculture in India supports directly or indirectly nearly 60% of the population. About 58% of the cultivated area in India (82.4 m ha) is rainfed which contributes 40% to the country's food production and is therefore particularly vulnerable to climate variability. As Indian agriculture is monsoon dependent, any temporal and spatial variations in rainfall will have reflective effect on water availability in both irrigated and rainfed areas, affecting the agriculture based economy. There are preliminary reports that the recent trend of decline in grain yields of rice and wheat in Indo-Gangetic Plains (IGP) could have been partly due to weather changes (Aggarwal *et al.*, 2004).

Climate change is now a reality and remains one of the biggest challenges facing society. Global climate change induced by increases in greenhouse gas concentrations is likely to increase temperatures, change precipitation patterns, including more total rainfalls and more frequent high intensity rainfall events. Changes in temperature and precipitation under climate variation have serious impacts on hydrologic processes and water resources availability, and thereby affecting agricultural production and productivity. Enhancing the resilience of Indian agriculture to combat climate change and climate variability is imperative to the livelihood security of farmers in the country. Assessment of climate change impacts and devising appropriate adaptation strategies will enable farmers to cope with various climate risks, promote efficient use of natural resources to bring sustainability to farm production and stability to their incomes.

Carbon capture and sequestration (CCS) can play a pivotal role in reducing India's emissions in the future. Land use systems that have higher carbon content than existing plant community can help achieve net gains in carbon,specifically and significant increases in carbon storage.This can be achieved by moving from lower biomass land uses [e.g. grasslands, crop fallows, etc] to tree based systems such as forests, plantation forests and agroforestry. Agroforestry provides a unique opportunityto combine the twin objectives of climate change adaptation and mitigation. Forestry has been recognized as a means to reduce CO_2 emissions as well as enhancing carbon sinks. The role of forests (or trees) in carbon cycles is well recognized and forests are a large sink of carbon. There is considerable interest to increase the carbon storage capacity of terrestrial vegetation through agroforestry.

Climate Change and Climatic Variability

Climate is the long-term average of a region's weather events lumped together. Climate change refers to "*a change in the state of climate that can be identified by changes in mean and/or variability of its properties and that persists for an extended period, typically decades or longer*". It is the change in climate over time whether due to natural variability or as a result of human activity (IPCC, 2007). Changes in land use pattern and concentration of greenhouse gases (GHGs) in atmosphere are two major anthropogenic causes of climate change and variation. Emissions of GHGs (carbon dioxide, methane, chlorofluorocarbons (CFCs) and nitrous oxide) are causing substantial increase in their concentrations in the atmosphere. This increase would enhance the greenhouse effect, resulting in additional warming of the earth's surface. The global atmospheric concentration of carbon dioxide has increased from a pre-industrial value of about 280 ppm to 379 ppm in 2005 (IPCC, 2007), and 395 ppm by volume as of August 2014. Similarly, anthropogenic nitrogen fixation has doubled as a result of increased fertilizer use, fossil fuel consumption and fixation by leguminous crops (Walker and Steffen, 1999). The atmospheric concentration of methane is reported to have increased at a rate of 1% per year (Glantz and Krenz, 1992) mainly due to wet-paddy rice farming and livestock farming. The global atmospheric concentration of methane has increased from a pre-industrial value of about 715 ppb to 1732 ppb in the early 1990s, and was 1774 ppb in 2005. Similarly, the global atmospheric nitrous oxide concentration increased from a pre-industrial value of about 270 ppb to 319 ppb in 2005 (IPCC, 2007).

Climate Change Projections

The Fourth Assessment Report (AR4) of Intergovernmental Panel on Climate Change (IPCC) projected an area-averaged increase in the median annual mean temperature in the range of 2.5 (South East Asia) to 4.3°C (Northern Asia) by the end of 21st Century (2080-2099) under A1b emission scenario (IPCC, 2007). The recent climate change projections (AR5), based on the Coupled Model Inter-comparison Project Phase 5 (CMIP5), projects a warming of 1.5, 2.4, 2.8 and 4.3 °C for India under the RCP2.6, RCP4.5, RCP6.0 and RCP8.5 scenarios, respectively, for 2080s (2071-2100) compared to the baseline period of 1961-1990 (Chaturvedi *et al.*, 2012). There is a consistent warming trend over the country in short, mid as well as long-term scenarios (Table 2.1). All-India annual mean ambient temperature increases by 1.7–2.0 °C during 2030s under different RCP scenarios and by about 2.0–4.8 °C during 2080s, relative to the pre-industrial base of 1880s (1861-1900). Precipitation is projected to increase almost all over India except for a few regions in short-term projections (2030s). The CMIP5 based ensemble projects an all-India precipitation increase of 6, 10, 9, and 14% increase under the RCP2.6, RCP4.5, RCP6.0 and RCP8.5 scenarios, respectively, for 2080s compared to the 1961-1990 baseline. All-India annual precipitation increases by 1.2–2.4% and 3.5–11.3% during 2030s and 2080s, respectively relative to the pre-industrial base under different RCP scenarios.

Table 2.1 CMIP5 model ensemble-based projected annual temperature change (°C) and precipitation change for 2030s, 2060s and 2080s relative to the pre-industrial baseline (1880s).

RCPs*	Temperature Change (°C)			Precipitation Change (%)		
	2030s	2060s	2080s	2030s	2060s	2080s
RCP2.6	1.7	1.9	2.0	1.2	2.7	3.5
RCP4.5	1.9	2.5	2.9	0.8	4.3	7.0
RCP6.0	1.7	2.4	3.3	1.6	3.5	6.8
RCP8.5	2.0	3.3	4.8	2.4	6.6	11.3

(Adapted from Chaturvedi *et al.*, 2012) *RCP: Representative Concentration Pathways

Climate Change Impact on Water Resources

The rising levels of CO_2 concentration and climate change may have significant effects on the hydrology and regional water resources availability. Under a doubling of CO_2 concentration, global mean runoff has been projected to increase by approximately 5% as a result of reduced ET due to CO_2 enrichment alone (Betts *et al.*, 2007). Several authors have carried out extensive studies in hydrologic impacts of climate change on Indian River basins using different hydrological models and climate change scenarios (Table 2.1). Simulation studies conducted for 12 river basins of India, using distributed hydrological model SWAT and HadRM2 regional climate model projected scenarios, indicated a general reduction in the quantity of the available runoff under the future (2041-2060) scenario (Gosain *et al.*, 2006). The river basin of Luni is expected to experience acute physical water scarcity; Mahi, Pennar, Sabarmati and Tapi are likely to experience constant water scarcity and shortage; Cauvery, Ganga, Narmda and Krishna are likely to experience seasonal or regular water stressed conditions; Godavari, Brahmani and Mahanadi are expected to experience water shortage only in few locations. However, when PRECIS RCM A1b scenario data was used along with the SWAT hydrological model, an increase in water yield (net amount of water leaving the sub-basin and contributing to the stream flow, and includes surface runoff, lateral flow and subsurface contribution) was reported in most of the Indian river systems during 2030s and 2050s, except in the Pennar basin where decrease in water yield during 2030s is more pronounced (Gosain *et al.*, 2011). Islam *et al.* (2012a) reported that the changes in temperature had a relatively lesser effect on the magnitude of annual and seasonal streamflow as compared to rainfall changes in the Brahmani River basin. A 10% decrease in rainfall resulted in 23% decrease in annual streamflow, whereas 4°C rise in temperature resulted in 11% decrease in annual streamflow. The combined effect of changes in rainfall and temperature changes showed changes in the mean annual streamflow in the range of -33 to 62%. A 4°C rise in temperature coupled with 30% increase in rainfall resulted in 62, 73, 39, 52 and 30% increase in annual, monsoon, pre-monsoon, post-monsoon and winter season streamflow, respectively. Similarly, a temperature rise of 4°C and a 10% decrease in rainfall resulted in 33, 35, 15, 32, and 21% decrease in annual, monsoon, pre-monsoon, post-monsoon and winter season streamflow, respectively.

Climate Change Impact on Irrigation Water Demand

Irrigation water demand is likely to be affected under the projected climate change scenarios due to changes in rainfall, temperature and the evaporative demand. The increase in temperature under the climate change scenarios is expected to increase the evapotranspiration demand. However, there are several studies indicating that the effect of rising temperature on ET is moderated by elevated CO_2 concentrations (Rosenberg *et al.*, 1989; Islam *et al.*, 2012b, 2012c), and thereby ET_o demand may not rise significantly. Simulation study conducted using meteorological data of Varanasi also demonstrated similar results (Priya *et al.*, 2014). They reported increase in reference ET_o by about 0.1 mm/d (2.3% increase with respect to baseline) annually with 1°C rise in mean temperature, and about 6% decrease in annual ET_o demand with doubling (660 ppm) of CO_2 concentration with temperature remaining constant. Simulating the combined effect of temperature and elevated CO_2 showed that the effect of about 1.0°C rise in temperature is offset by increase in CO_2 levels upto 495 ppm, and 2.5°C rise in temperature is offset by increase in CO_2 levels upto 660 ppm.

Climate Change Impact on Agriculture

Agricultural productivity is sensitive to two broad classes of climate-induced effects, one is the direct effects due to changes in temperature, precipitation, and carbon dioxide concentrations and the other is the indirect effects through changes in soil moisture and the distribution and frequency of infestation by pests and diseases (Mendelsohn 2014). The main direct effect is generally seen on the duration of the crop. An increase in temperature will speed up crop development. In the case of an annual crop, the duration between sowing and harvesting will shorten, for example, the duration in order to harvest a maize crop could shorten between one and four weeks. The shortening of such a cycle could have an adverse effect on productivity because senescence would occur sooner. In India impact of 1-2^{0}C increase in mean air temperature is expected to decrease rice yield by about 0.75 t/ha in efficient zones and 0.06 t/ha in coastal regions and impact of 0.5^{0}C increase in winter temperature is projected to reduce wheat yields by 0.45 t/ha.

Furthermore, crops may experience both low and high weather extremes like drought and flood, heat and chilling etc. in a single cropping season and such changes will have varying and complex impacts on agricultural production. Reductions in yields as a result of climate change are predicted to be more pronounced for rainfed crops in comparison with irrigated crop and under limited water supply situations because there are no coping mechanisms for rainfall variability (Dasgupta *et al.*, 2013). Additional challenge to temperature increase stems from the fact that higher temperatures will increase the rate at which plants lose moisture resulting in increased transpiration and water loss. Agricultural production in India will be adversely affected not only by an increase or decrease in the overall amount of rainfall, but also by shifts in the timing of the rainfall.

In India the impact of climate change on agriculture is expected to be more severe than realized earlier, in particularly in wheat if there is one degree rise in temperature in areas like Uttar Pradesh, Punjab and Haryana could amount to a loss of about several million tonnes annually, in addition production in crops such as rice, millet and maize could reduce upto 10 percent. Projected impacts of climate change on Indian Agriculture are summarized below:

- Increase in CO_2 to 550 ppm increases yields of rice, wheat, legumes and oilseeds by 10-20%.
- A 1^0C increase in temperature may reduce yields of wheat, soybean, mustard, groundnut and potato by 3-7%. Much higher losses at higher temperatures.
- Productivity of most crops to decrease marginally by 2020 but by 10-40% by 2100.
- Significant impacts on food quality- basmati rice, wheat, etc.
- Possibly some improvements in yields of chickpea, rabi maize, sorghum and millets; and coconut in west coast.
- Less loss in potato, mustard and vegetables in North-Western India due to reduced frost damage.
- Droughts and floods are likely to increase production variability.
- Considerable effects on microbes, pathogens, and insects.
- Increasing sea and river water temperatures are likely to affect fish breeding, migration and harvests.

Results of simulation studies quantifying the impacts of climate change on for the major crops are summarized in Table 2.2.

Adaptation to Climate Change

Climate change is expected to produce water stresses in several parts of the country, and thereby affecting crop production. Substantial adaptation efforts will be needed to ensure adequate supply and efficient utilization of available water resource.There is great potential for short-term adaptation is in demand control and more efficient and integrated management of surface and groundwater supplies. Adoption of drip/ micro-irrigation, rainwater harvesting, groundwater recharge, encouraging water-saving techniques such as water recycling, reducing losses in canal systems could form some of the adaptation options. Participatory integrated watershed management approach covering biophysical and socio-economic interventions has shown rich dividends at ensuring resilience to climate change. In the agriculture sector, the adaptive measures to counter detrimental impacts of climate variations include changing planting/sowing dates, changing crops and crop varieties that are more tolerant to climatic variations.Adopting improved varieties, efficient application and re-scheduling of inputs to suit crop phenology can reduce yield loss in future climate. Further, application of higher dose of fertilizers (about 25% higher than amount currently applied by farmers) can increase all India irrigated rice yields by about 15, 12 and 8% over current values during 2020s, 2050s and 2080s scenarios, respectively.

Table 2.2 Climate change impact hydrology and water resources in India

River basin	Impact
The Ganges basin	Increases in mean annual runoff in all the sub basins
Spiti basin	Under a warmer climate scenario, snowmelt runoff and glacier melt runoff cause an earlier response of total streamflow and a change in flow distribution Increase in temperature resulted in a large increase in streamflow during the pre-monsoon season followed by monsoon season.
River basins of Central India	The Sher basin is more sensitive to climate change followed by Kolar and Damanaganga
Kosi basin	A 2% to 8% decrease in runoff with 4°C temperature rise and no change in precipitation.
Damodar River basin	Reduced water availability for the 2050s and 2060s scenarios.
Satluj River basin	Snowfed basins are more sensitive in terms of reduction in water availability
River basins of India*	A general reduction in the quantity of the available runoff under the future (2041-2060) scenario.

River basins of India**	An increase in water yield in most of the Indian river systems during 2030s and 2050s
Mahanadi River basin	A decreasing trend in future monsoon streamflow; Increase in probability of extreme, severe, and moderate droughts; expected to experience progressively increasing intensities of flood in September and drought in April
The Brahmaputra River	Increase in peak discharge and flood duration during pre-monsoonal and monsoonal seasons, but reduction in number of flood waves per season.
Tunga–Bhadra river basin	Increase in average annual streamflow during 2020s, 2050s, and 2080s
Brahmani River basin	Streamflow is more sensitive changes in rainfall as compared to temperature changes; Increase in annual as well as seasonal streamflow in the basin
Kangsabati river catchment	Decrease in reservoir inflows

Source: Islam et al. (2014), *Brahmani, Cauvery, Ganga, Godavari, Krishna, Luni, Mahanadi, Mahi, Narmada, Pennar, Sabarmati, Tapi **Baitarni, Brahmani, Brahmaputra, Cauvery, Ganga, Godavari, Indus, Krishna, Luni, Mahanadi, Mahi, Meghna, Narmada, Pennar, Sabarmati, Subernrekha, Tapi

Table 2.3 Impact of climate change on crop production in India

Crop	Study area	Result
Wheat	72 locations spread across wheat growing region	A two degree increase in temperature will reduce potential yields but will have small effect on irrigated yields in northern India. The effect of climate change will be more pronounced in central India.
Wheat	India	Decrease in wheat yields under the projected climate scenarios of a global and a regional climate model.
Rice and Wheat	North-West India	A 3 and 2°C rise in air temperature offsets the positive effect of elevated CO_2 on rice and wheat yields, respectively.
Rice and Wheat	Ganga basin (India)	Reduction of potential rice and wheat yield for both the scenarios. In lower Ganga basin, HadRM3 projections showed increase in the potential rice and wheat yields whereas,

		REMO RCM projections showed decreased potential yields.
Rice	Kerala, India	Increase in rice yields under rainfed conditions.
Rice	4 sites covering different agro-climatic regions	Increase in rice yields in pessimistic as well as optimistic scenarios of climate change.
Rice	India	Decrease in irrigated and rainfed rice yield, but decrease in rainfed rice yield is marginal during 2050s and 2080s.
Maize	Indo-Gangetic plains	Decrease in winter maize grain yield in mid Indo-Gangetic plains and southern plateau with an increase in temperature. In the upper Indo-Gangetic plain (UIGP), yield is projected to increase up to a 2.7°C rise in temperatures.

Source: Islam *et al.* (2014)

Carbon Sequestration in Agriculture

Carbon sequestration in agriculture sector refers to the capacity of agriculture lands and forests to remove carbon dioxide from the atmosphere. Carbon dioxide is absorbed by trees, plants and crops through photosynthesis and stored as carbon in biomass in tree trunks, branches, foliage and roots and soils (EPA, 2008b). Forests and stable grasslands are referred to as carbon sinks because they can store large amounts of carbon in their vegetation and root systems for long periods of time. Soils are the largest terrestrial sink for carbon on the planet. The ability of agriculture lands to store or sequester carbon depends on several factors, including climate, soil type, type of crop or vegetation cover and management practices.

The amount of carbon stored in soil organic matter is influenced by the addition of carbon from dead plant material and carbon losses from respiration, the decomposition process and both natural and human disturbance of the soil. By employing farming practices that involve minimal disturbance of the soil and encourage carbon sequestration, farmers may be able to slow or even reverse the loss of carbon from their fields.

Scope for C economy in Indian Agriculture

The C economy in Indian agriculture can be improved through two principal modes by reducing the total emission, and (2) by increasing the C sink.

Enteric fermentation: The emissions from agriculture sector are dominated by the emissions from enteric fermentation from livestock, that contributes about 63% of the total emissions from the agriculture sector (INCCA, 2010). Most of the CH_4 from ruminant livestock originates from microbial fermentation of carbohydrates in the rumen and the lower digestive tract, referred to as enteric CH_4 emission. Methane mitigation strategies can be broadly divided into preventive and 'end of pipe' options. *Preventive measures* reduce C/nitrogen inputs into the animal husbandry, generally through dietary manipulation. Alternatively, *'end of pipe'* options reduces or inhibit the production of CH_4 (methanogenesis) within the animal husbandry.

Manure management: Management decisions about manure disposal and storage affect emissions of CH_4 and N_2O, that are formed during decomposing manures. Although CH_4 and N_2O emissions from manure management are minor, manure itself is an important contributor to emissions because it is either applied on cropland as organic fertilizer or directly deposited by grazing animals in pastures. Biogas generation from the cattle dung reduces the net emission under the particular area.

Rice cultivation: The acreage under rice cultivation in India is estimated at about 44.0 million hectares, mainly planted in wet monsoon or irrigated systems by flooding and puddling the fields. The rice fields are a major source of emission of GHGs as CH_4 and N_2O.

Soil C sequestration: Globally, potential for C sequestration in soils over the next 50 year period is estimated at 2 to 3 billion tons C yr^{-1} through improved management of existing agricultural soils and restoration of degraded lands (Lal, 2009).The potential of soil C sequestration in India is estimated at 39 to 52 million tons C yr^{-1} (Lal, 2004), that includes restoration of degraded soils (7.2-9.8 million t C yr^{-1}) and reduction in erosion-induced emission of C (4.3-7.2 million tons C yr^{-1}) (Lal, 2004).

Field burning of crop residues in field: Indian agriculture produces about 500-550 million tons (Mt) of crop residues annually. However, a large portion of this, about 90-140 Mt annually is burnt on-farm primarily to clear the field for sowing of the succeeding crop.

Benefits of C-economy: It is estimated that 24 to 40 million tons of additional food grains can be produced annually if SOC pool in soils of the developing countries can be enhanced @1 t/ha/year (Lal, 2006).

As per the study in dry land soils of India, every t ha^{-1} increase in SOC stock in the root zone, an increase in grain yield (kg ha^{-1}) of 13 for groundnut, 101 for finger millet, 90 for sorghum, 170 for pearl millet, 145 for soybean, 18 for lentil, and 160 for rice (Srinivasarao *et al.*, 2013) was observed.

Each ton of sequestered organic C in the plough layer has been reported to compensate for 4.7 kg fertilizer N ha^{-1} in irrigated wheat (Benbi and Chand, 2007). This indirectly results in reducing emissions related to fertilizer N production and application.

Strategies for increasing Soil C sequestration

- To reduce water erosion and thus minimizing the loss of soil and soil C.
- Optimum and balanced plant nutrition through integrated nutrient management (INM) methods, which should be soil, crop and region specific.
- Restrictions (either through legal prohibition or incentivization) on crop biomass and residue burning.
- Promotion of conservation agriculture and minimum tillage methods.
- Creation of model villages for C economy to demonstrate an array of activities including biogas use, vermicomposting, on-farm biomass generation, tank silt application in crop fields, legume intercropping, bund planting with green manuring plants etc.
- Maintaining a minimum soil C level by regular additions of organic inputs.
- Promotion of Agroforestry systems

Forest biomass carbon in eastern region of India

The eastern region comprised of states like, Assam, Bihar, Jharkhand, West Bengal, Odisha, Chhattisgarh. The eastern states cover 19.24 Mha of forest and tree cover with highest area of 5.95 Mha in Chhattisgarh. The total growing stock of forest and tree outside forest accounted for 1037.075 and 329.225 million m³, respectively. The highest growing stock in the eastern region was 1366.3 million m³.

Table 2.4 Forest area and growing stock in different eastern states

States	Forest and tree cover (km²)	Percent GA	Growing stock (million m³)		
			Forest	Tree outside forest (TOF)	Total
Assam	29, 237 (2.92 Mha)	37.27	173.494	41.336	214.830
Bihar	9, 214 (0.92 Mha)	9.79	35.186	47.195	82.381
Chhattisgarh	59, 540 (5.95 Mha)	44.04	334.381	70.069	404.450
Jharkhand	25, 891 (2.58 Mha)	32.48	116.308	51.308	167.616
Odisha	53, 204 (5.32 Mha)	34.17	285.191	73.624	358.815
West Bengal	15, 330 (1.53 Mha)	17.27	92.515	45.693	138.208
Total	1, 92, 416 (19.24 Mha)	1037.075	329.225	1366.3	

Source: FSI, 2011; 10000 km²= 1 Mha (million ha)

Agroforestry systems for increasing C-sequestration

Agroforestry enhances carbon uptake by lengthening the growing season, expanding the niches from which water and soil nutrients are drawn and, in the case of nitrogen (N)-fixing species, enhancing soil fertility. The result is that when agroforestry systems are introduced in suitable locations, carbon is sequestered in the tree biomass and tends to be sequestered in the soil as well. Improved management in existing agroforestry systems could sequester 0.012 Tg C yr^{-1} while conversion of 630 million ha of unproductive or degraded croplands and grasslands to agroforestry could sequester as much as 0.59 Tg C annually by 2040 (IPCC, 2000), which would be accompanied by modest increases in N_2O emissions.

Table 2.5 Forest Biomass Carbon in Eastern India (Million tonnes)

Items	Factor	Assam	Bihar	Chhattisgarh	Jharkhand	Odisha	West Bengal	Total
Growing stock, GS (Mm^3)		214.83	82.38	404.45	167.62	358.82	138.21	1366.30
Mean biomass expansion factor (BEF)	1.575							
Ratio Below to Above ground biomass (RBA)	0.266							
Above ground biomass, AGB (Volume, m^3) (=GS x BEF)		338.36	129.75	637.01	264.00	565.13	217.68	2151.92
Below ground biomass, BGB (Volume, m^3) (=AGB x RBA)		90.00	34.51	169.44	70.22	150.33	57.90	572.41
Total biomass, TB (Volume) (=AGB + BGB)		428.36	164.26	806.45	334.22	715.46	275.58	2724.33
Mean density, MD	0.7116							
Biomass (Mt), B (=TB x MD)		304.82	116.89	573.87	237.83	509.12	196.10	1938.64
Ratio (Other Forest Floor Biomass except tree to Tree Biomass), ROFFB	0.015							
Biomass (other vegetation), (=B x ROFFB)		4.57	1.75	8.61	3.57	7.64	2.94	29.08
Total Forest Biomass in Mt (Trees + Shrubs + Herbs)		309.39	118.64	582.48	241.40	516.76	199.04	1967.72
Dry Weight in Mt (80% of TFB)		247.51	94.91	465.98	193.12	413.41	159.24	1574.17
Carbon in Mt (40 % of DW)		99.01	37.97	186.39	77.25	165.36	63.69	629.67

Different factors for calculation was derived from Kaul *et al.*, 2009; Roy and Ravan, 1996 and Kiswan *et al.*, 2009.

In an agri-silvicultural system, *Dalbergia sissoo* at age 11 years was able to accumulate 48-52 t/ha of biomass (Newaj and Dhyani, 2008). Carbon dynamics involving different pruning treatments were studied in an agri-silvicultural system where tree biomass was 23.61 to 34.49 tC/ha with black gram-mustard. In a study on poplar based agri-silvicultural system, total biomass in the system was 25.2 t/ha, 113.6% higher than sole wheat cultivation, where net carbon storage was 34.61 tC/ha compared to 18.74 tC/ha in sole wheat cultivation (Chouhan *et al.*, 2010). In a system comprising *Albizia* and mixed tree species like Mandarin accumulated 1.3 t biomass/ha storing 6939 kg/ha in tree and crop biomass was reported (Sharma *et al.*, 1995).

Soil organic carbon enhancement: In a study conducted on intercropping of trees with crops, Singh *et al.* (1989) reported an enhancement in SOC by 33.3 to 83.3% for *Populus deltoides* and *Eucalyptus* hybrid with *Cymbopogon* sp., with a greater increase in SOC under *Populus deltoides* plantation. Soil organic carbon has been reported to have improved for agroforestry plantations ranging in age from 6 years (Maikhuri *et al.*, 2000) to 20 years (Saha *et al.*, 2007).

In a Poplar based agroforestry system, trees could sequester higher soil organic carbon up to 30 cm depth during the first year of plantation (6.07 t/ha/yr) than in subsequent years (1.95-2.63 t/ha/yr) with greater soil carbon storage in sandy clays than loamy sand (Gupta et al., 2009). Traditional *Prosopis cineraria* based systems lead to a 50% increase in SOC largely due to leaf litter (Venkateswaralu, 2010). Samra and Singh (2000) observed an increase in soil organic carbon status of surface soil by 0.39 to 0.52% under *Acacia nilotica+Sacchram munja* and 0.44 to 0.55% under *Acacia nilotica+Eulaliopsis binata* after saccarum 5 years.

1. Silvipastoral systems

Carbon sequestration in tree biomass: Comparative studies conducted by NRCAF (2007) on biomass production from natural grassland and silvipastoral system comprising *Albizia amara, Dichrostachys cinerea* and *Leucaena leucocephala*as woody perennials with *Chrysopogan fulvus* as grass and *Stylosanthes hamata* and *S. scabra*as legume revealed that in 8 years, rate of biomass carbon stored in silvipastoral system was 6.72 tC/ha/yr, two times more than 3.14 tC/ha/yr from natural grassland. Singh (2003) estimated the total carbon sequestered in farm forestry with species such as *Eucalyptus sp., Populus deltoides, Tectona grandis,*

*Anthocephalus chinensis*trees to be around 16,400 t/yr. Rai *et al.* (2009) studied the effect of introducing a silvipastoral system in a natural grassland in semi arid Uttar Pradesh, where introduced species of *Albizia procera, Eucalyptus tereticornis, Albizia lebbeck, Embilica officinalis* and *Dalbergia sissoo* accumulated 8.6, 6.92, 6.52, 6.25 and 5.41 t/ha/yr of biomass. Here, the carbon storage in the system was 1.89-3.45 tC/ha in silvipasture and 3.94 tC/ha in pure pasture. Saharan et al. (2002) reported an increase in organic carbon of 1.7 to 2.3 times in a silvipastoral system involving *Leucaena leucocephala, Cenchrus ciliaris* and *Stylosanthes hamata* compared to a control.

Soil organic carbon enhancement: Narain (2008) reported that planting trees and grasses in degraded lands in arid areas can help increase soil carbon stock from 24.3Pg to 34.9 Pg. Studies conducted at RRS, Kukma have reported a total belowground carbon stock of 1.6 t/ha, 23.4% of the total carbon stock under silvipastoral system involving *Acacia tortilis*and *Cenchrus setegerus* (Kumar, 2010).

2. Carbon stored in block and boundary plantations

In a study done by Kumar (2010) on four different agroforestry systems (*Populus deltoides* block plantation+wheat, *Eucalyptus* hybrid boundary plantation+wheat, *Populus deltoides* boundary plantation+ wheat and *Populus deltoides* block plantation+lemon grass), all 9-year old block and boundary plantations, it was observed that total carbon sequestration [in trees] was 70.59, 21.38, 116.29 and 18.53 t C ha^{-1} in system *Populus deltoides*+wheat followed by 68.53, 20.63, 113.03 and 17.60 t Cha-1 in system *Populus deltoides*+Lemon grass, respectively, with a greater potential for carbon sequestration in boundary plantations of *Populus deltoides*and *Eucalyptus* hybrid. In a study on mitigation potential of block plantations in farm forestry, Hooda et al. (2007) estimated a mitigation potential of 48.5, 62.7, 61.7, 60.8, 37.6 tC/ha/yr respectively for Khair, Chir pine, mixed plantations, Mango and Kinoo based farm forestry systems in Uttarakhand.

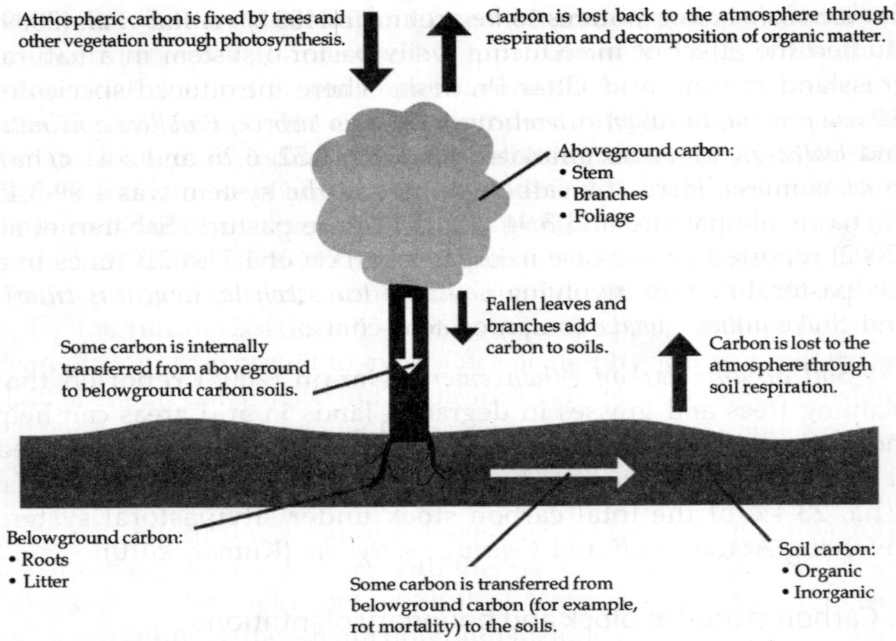

Fig. 2.1 Carbon pools in forestry and agriculture (Source: EPA)

Agricultural production systems for increasing C-sequestration

Several farming practices and technologies can reduce greenhouse gas emissions and prevent climate change by enhancing carbon storage in soils; preserving existing soil carbon; and reducing carbon dioxide, methane and nitrous oxide emissions.

1. Conservation tillage and cover crops

Conservation tillage refers to a number of strategies and techniques for establishing crops in the residue of previous crops, which are purposely left on the soil surface. Reducing tillage reduces soil disturbance and helps mitigate the release of soil carbon into the atmosphere. Conservation tillage also improves the carbon sequestration capacity of the soil. Additional benefits of conservation tillage include improved water conservation, reduced soil erosion, reduced fuel consumption, reduced compaction, increased planting and harvesting flexibility, reduced labour requirements and improved soil tilth.

2. Improved cropping and organic systems

Organic systems of production increase soil organic matter levels through the use of composted animal manures and cover crops. Organic cropping systems also eliminate the emissions of greenhouse gases from the production and transportation of synthetic fertilizers. Components of organic agriculture could be implemented with other sustainable farming systems, such as conservation tillage, to further increase climate change mitigation potential.

Generally, conservation farming practices that conserve moisture, improve yield potential and reduce erosion and fuel costs also increase soil carbon.

Examples of practices that reduce carbon dioxide emissions and increase soil carbon include direct seeding, field windbreaks, rotational grazing, perennial forage crops, reduced summer fallow and proper straw management. Using higher-yielding crops or varieties and maximizing yield potential can also increase soil carbon.

Tillage methods that control weed without damaging the soil and losing carbon often preserves soil carbon.

3. Use of microorganisms to convert soil carbon into stable forms

It is essential to manage the soil to maintain high levels of soil organisms. Also it is essential that farming techniques stimulate the species of soil microorganisms that create stable carbons, rather than convert it into CO_2. The stable farms of soil carbon such as humus and glomalin are manufactured by microorganisms. They convert the carbon compounds that are readily oxidised into CO_2 into stable polymers that can last thousands of years in the soil. Stimulating the carbon builders in soil rather than carbon eaters, can help in maintaining soil organic carbon.

4. Reduce nitrogen applications

Synthetic N fertilizers are one of the major causes of the decline of soil carbon. This is because it stimulates a range of bacteria that feed on nitrogen and carbon to form amino acids for their growth and reproduction. Every ton of nitrogen applied results in the bacteria consuming 30 tons of carbon.

5. Land restoration and land use changes

Land restoration and land use changes that encourage the conservation and improvement of oil, water and air quality typically reduce greenhouse gas emissions. Modifications to grazing practices, such as implementing sustainable stocking rates, rotational grazing and seasonal use of rangeland, can lead to greenhouse gas reductions. Converting marginal cropland to trees or grass maximizes carbon storage on land that is less suitable for crops.

6. Methane capture

Large emissions of methane and nitrous oxide are attributable to livestock waste treatment, especially in dairies. Anaerobic digestion converts animal waste to energy by capturing methane and preventing it from being released into the atmosphere. The captured methane can be used to fuel a variety of on-farm applications, as well as to generate electricity. Additional benefits include reducing odors from livestock manure and reducing labor costs associated with manure removal.

□□□

3

Water Harvesting and Management

Per capita water availability of many river basins in India is declining over the years due to sustained population pressure, agriculture and industrial expansion, besides changing climate scenarios. This is particularly evident from the fact that the per capita availability has decreased from 1816 m^3/year in 2001 to 1545m^3/year in 2011 (Jha, 2013). Particularly rainfed areas, which produce about 60% of the total food produced in India, face recurrent water shortages during dry seasons. Much of this land is located in the arid or semi-arid belts where rain falls irregularly. Recent droughts due to delay or failure of monsoon has highlighted the potential risks to crops, humans and livestock. Provision of irrigation in such areas will be an ideal response to draught. High initial costs of the systems and with increasing fuel cost (diesel, petrol or electricity), the irrigation is available with few who can afford to dig tube wells or invest on lift irrigation. There is now increasing interest in low cost alternatives like "water harvesting". Water harvesting offers many opportunities to increase the farm income by providing irrigation to crops and through culturing fish and duck in the stored water. The cost of irrigation is reduced as the water is available nearby the area of its utilization. Water harvesting can be combined with groundwater recharge schemes to improve the water table.

Eastern Plateau and Hill Region (EPHR) is an important agricultural region of the country covering 64 districts of Chattisgarh, Jharkhand, Odisha and some parts of Maharashtra and Madhya Pradesh. It is characterized by semi humid climatic conditions and gets an average annual rainfall of about 1350 mm received within an average of 130 rainy days in a year. The state receives about 10 million hectare meter rainfall, out of which about 20 % is lost in the atmosphere, 50 % flow as surface runoff and balance 30% soaks into the ground as soil moisture and ground water. Despite of high rainfall the region faces water shortage during dry season. Lack of water harvesting and poor rain water management leads to reduced water availability during the non-monsoon period. Paddy is the main kharif crop which is prone to vagaries of the monsoon season. It is highly prone to draught that combines with frequent and lengthy dry spells during the monsoon season. Very small areas are put under crops like pulses, oilseeds and maize during Rabi season while most of the land is kept fallow during the summer season. This is mainly because of lack of sufficient water to irrigate the crops during this period.

The water shortage is linked directly to the undulating topography of the region which allows little opportunity time for the water to infiltrate and join groundwater. The delayed subsurface flows have little contribution in the main stream flows resulting in drying up of the streams immediately after withdrawal of the monsoon. The hydrograph of the main streams is of typical ephemeral type with high peaks and short base. Under this type of hydrological setup, substantial amount of generated runoff quickly reaches the watershed outlets making rainwater unavailable for the crop production or other activities in the watershed.

Rainwater harvesting is a technique used for collecting, storing and using rainwater for domestic, agricultural or any other uses. The rainwater is collected from various hard surfaces such as rooftops, runoff from catchments, from streams and water conservation through watershed management or other manmade aboveground hard surfaces. It is an age-old system of collection of rainwater for future use. The harvested water can be stored on surface through ponds and tanks or can be recharged to groundwater. With implementation of water harvesting structures at farm levels, the water availability for the agricultural production can be increased. Runoff from the uplands can be harvested at the mid lands to irrigate the mid and low lands of the region. If electricity is available to operate the irrigation pumps, the harvested water can be even used to irrigate the uplands during lean season. Lowlands have fairly good groundwater supply where groundwater can be harvested and used in agriculture. However, pumping water from low to uplands may increase the cost of irrigation. Harvesting rain water in the uplands for the providing lifesaving irrigation to the newly established fruit/forestry plants can be a better alternative. Retention of water in the uplands and midlands through unlined water harvesting structure improves groundwater recharge. Success of rainfed agriculture can be ensured with appropriate utilization of every drop of water.

Water Source

Runoff generated from the lands, perennial streams, natural springs, Nalas and sometime waste water from well and municipal hand pumps can be the source of water for water harvesting structures. There are other potential artificial sources of water which include;

- Collecting water from the roof (due to rain) of residential areas using *"patnala"* (thin iron sheet bent in the shape of English alphabets "U" and "V") and storing the harvested water into tanks through PVC pipes and filters.
- Collecting surface runoff generated from rain by formation of ridges and gully on land surfaces and finally channelizing into the tanks.
- Diverting water from natural sources such as perennial streams, springs and *"Nalas"*.
- Collecting waste water of well/hand pump etc.

Storage Structures

In the high altitude areas, the river flow in deep valley at the toe of slopes rarely serve any purpose as far as domestic water supply and irrigation are concerned. This has forced the agriculture in the high altitude areas to be rainfed. Despite of having high annual rainfall, its distinct topographical features and steep terrains provide little opportunity time for rainwater to infiltrate. It has been estimated that only less than 15 per cent of the rainwater is able to percolate down through slopes and recharge the soil moisture. The rainfall pattern in the hill is erratic i.e. most of rain (approximately 70% of annual rain) received only during the two months of monsoon. Furthermore, the climate of high altitudes comprises several climatic anomalies such as wide variability in temperature (-3^{0}C to 35^{0}C), frequent snowfall, hail, sleet and erratic pattern of rainfall. Under such abrupt conditions crops grown in open conditions could not be sustained.

Storage of harvested water is essential under these circumstances. The storage structures are designed to meet the requirement of different sectors. It should have good bottom and side sealing to check the seepage loss. Depending the site conditions and the purpose of storage, following structures are commonly used for storage of water for domestic as well as irrigation purpose in hilly areas.

- *Kachha* tank.
- Cemented (RCC) tank.
- Poly-tank or LDPE (Low Density Polyethylene) tanks.
- Syntax/ferro-cement tank.

Poly-Tanks in Hills

In hills, agriculture is main source of livelihood of rural people. However, small and scattered land holding, difficult topography, women dependent agriculture, totally rainfed, typical socio-economic factors and migration of youth towards urban/plain areas are major constraints of hill horticulture. In present scenario, concept of poly-tank cum micro-irrigation system can be employed in protected cultivation. Such system is best way to provide irrigation, under low water availability areas with high water use efficiency. The micro-irrigation system saves water (40-50%), fertilizer (25-30%), labour expenditure (50%), soil-erosion cum fertility (90%), minimizes plant mortality (80%), enhances crop yield (25-75 %) and improves the quality of produce as compared to traditional pattern specially when employed under protected environment.

Owing to distinct topology, soil characteristics and rainfall pattern, Low Density Poly Ethylene (LDPE) lined tanks are more suitable for the hilly regions. Research conducted at Krishi Vigyan Kendra, Lohaghat have highlighted several advantages of poly tanks over other water harvesting structures;

- Majority of hill farmers are small and marginal with scattered land holding who cannot afford to construct cement tank. Poly-tanks are low cost structures as compared to cemented tanks and plastic tanks. The cost of construction of cement tank is typically 5-6 times more as compared to similar capacity a poly-tank.

- Landslides and earthquakes are common features of the hilly region and risk of cracking of cemented tanks is very high. Poly-tanks adjust their shape according to minor disturbance in land due to their flexible nature and are ideal under this kind of setup.
- Huge loss of water is involved in long term storage of water in "*kachha*" tank. Lining of tank with LDPE film provides better option to reduce the seepage losses from the storage.
- Use of plastic tanks/ferro-cement tanks is restricted due to their limited capacity and handling problem.
- Poly-tanks are simple to construct, easy to maintain and requires little technical knowledge.

Selection of site: Selection of site is one of the important steps in construction of a poly-tank. The following points should be kept in mind during selection of site:

- Site should be at a place where maximum amount of rain water can be collected. It is better to construct near any building so that water from roof of the building can be harvested and easily stored in the poly-tank.
- Site should be as flat as possible and should not be prone to erosion.
- Poly-tank should be constructed at least 2 m above the irrigated area so that sufficient head for drip irrigation system can be easily achieved through gravity.
- Site should be out of reach from animals and children.

Estimation of capacity of Poly-tank: Pond with higher capacity will incur added investment while a lower capacity tank may not fulfill the requirement during dry season. Therefore, designing a pond of appropriate capacity is important. Calculation for the capacity of tank required for giving 10 cm irrigation in 1 *nali* (200 m^2) area is as follows:

$$\text{Capacity of poly-tank (V)} = \text{Depth of irrigation (m)} \times \text{Area } (m^2)$$
$$= 0.10 \text{ m} \times 200 \text{ m}^2$$
$$= 20 \text{ m}^3 \text{ or } 20{,}000 \text{ liters}$$

Shape of the Poly-tank: Water can be stored in tanks having square, rectangular and circular shape. Circular tanks can store more volume than the other shape for same wetted area which is advantageous in reducing evaporation and deep percolation losses. But the construction of circular is technically tough and square or rectangular shaped poly-tanks are generally preferred.

Trench Opening Poly tank

Fig. 3.1 Constructionof poly tanks

Length and width of the tank of the selected capacity can be worked out using following formula for rectangular shape tank.

$$V = \frac{D}{2}\{LW + (L - 2D)(W - 2D)\} \times 1000$$

Where,

V = capacity or volume of tank (liter)

D =depth of tank (meter)

L = length of tank (meter)

W = width of tank (meter)

It is worth notable that a safety factor of 10% in required volume should be taken during the design of poly-tank. Accordingly the sides of 20,000 liter capacity tank (1:1 side slope), would be 5.1m (L) x 5m (W) at top, 2.1m (L) x 2m (W) at bottom with a depth of 1.5 m.

Slope of walls of Poly-tank: Adequate slope of the side wall of the tank is very important factor in construction of poly-tank. The main purpose of providing slope in the side wall is to divert the pressure of water through the side wall of the tank unless polythene sheet may be punctured or tear. Normally, the slope of side wall is kept 1:1.

Depth of Poly-tank: In hilly areas soil depth is comparatively less, so depth of tank should not be more than 1.5 m. Construction of tanks with higher depths is tough and expensive.

Type of Polythene Sheet: Life of a poly-tank largely depends upon the quality and thickness of the polythene sheet. Generally, polythene sheet is available in black, blue, white and other colours in market. Its thickness is measured in GSM, micron (μ) and sometime in gauge. A sheet having specification of 250 GSM means the weight of 1 square meter piece of this polythene sheet will be 250 gram or in other words 1 kg polythene sheet of 250 GSM specifications will cover 4 square meter area. Maximum size of polythene sheet is available in 50 m x 7 m dimensions in the market. A poly-tank of maximum 5 m width can be constructed by this sheet. However, the tanks larger than this width can be constructed by sealing two polythene sheets using heat sealer (joining two sheets at high temperature). Polythene sheet of desired width can be obtained directly from the manufacturing company on demand. Size of polythene (1:1 slope tank) can be measured by the following formulae:

$$\text{Total length of polythene (m)} = 2D\sqrt{2} + (L - 2D) + 1.5$$

$$\text{Total width of polythene (m)} = 2D\sqrt{2} + (W - 2D) + 1.5$$

Procedure of poly-tank construction: Construction of poly-tanks requires considerations of several factors related to site and capacity of the tank. Step by step procedure of poly-tanks construction is provided here;

- The site for constructing poly-tank should be cleaned firstly so that the size of tank can be marked on the ground by lime. Once the rectangle/square having side equal to side of top of poly-tank marked, another rectangle/square will be marked inside this. Notable that side of inside rectangle/square will be less than the outer rectangle/square by twice the depth of tank if side wall slope is 1:1.
- Now digging will be started straight downward through the sides of inside rectangle/square up to the depth of the tank i.e. 1.5 m taken in this example. Now join the sides of outer rectangle with the bottom side of inner rectangle by cutting the soil in sloping manner. Finally we will get a tank having 1:1 side slope.

- All the stone, gravels, root and residue of vegetation should be removed from all surface of tank because these may puncture the polythene sheet. A spray of weedicides like Atrazine @ 0.4 gram/m^2 is recommended to avoid the growth of any kind of vegetation from the wall of the tank.
- Now, a mixture of clay, dung and straw is prepared and pasted on the side walls and bottom of the tank so that all these surfaces can be smoothened. Ultimate aim to stop all those chances due to which polythene sheet can be punctured.
- Now the polythene sheet of required size, calculated from the formulae discussed above, will be fitted in the excavated tank. The polythene sheet will be well covered by stone/brick and soil from all four upper corners.
- Pitching of boulders, tar-felt sheet, and some time bricks are recommended for long life of the polythene but it is not economical from framer point of view. However, now a day's good quality of polythene is available in the market which is resistant from sunlight, air, temperature variation etc.
- A small pit (1mx1mx1m) is recommended at the entry point of water so that soil and waste material with surface runoff should not enter into main tank.

Table 3.1 Cost of construction of a 20,000 liter capacity poly-tank

S.No.	Particular	Quantity	Rate* (Rs/unit)	Amount (Rs)
1.	Digging of tank	22.3 m^3	50	1115.00
2.	Pasting of mixture of clay, dung and straw	1 job	One time	500.00
3.	Polythene sheet (250GSM)	54.4 m^2	75	4080.00
4.	Spray of Atrazine inside tank	1 job	One time	100.00
5.	Miscellaneous charges (@4%)			232.00
	Total Cost (Rs)			**6027.00**

*Rate quoted in the table is based on the prevailing market price of Rudrapur, Uttarakhand and subject to vary place to place.

Polythene Lined Ponds

Undulating topography with its three distinct landforms viz uplands, midlands and low lands can be advantageously used to harvest, store and utilize the rainwater. The potential of this kind of topography can be tapped through devising water harvesting structures that make the use of uplands as source of water and lower midlands or midlands as command area where the harvested water can be efficiently used. In midlands and lower uplands polythene lined ponds of suitable size can be constructed to collect and store the runoff.

Location of the ponds is selected such that there is enough catchment area to ensure water pond supply to its full capacity. The catchment area can be planted with tree based plantations having minimum irrigation water requirement. The area needs to be protected from erosion by providing proper grass cover and/or erosion control structures if required. Ploughing of catchment area should be avoided to ensure runoff generation. If necessary some grassed water ways can be built up to direct the flow of water towards the inlet of pond. The command area can be planted with fruit crops. The interspaces can be utilized to cultivate crops like groundnut, brinjal, tomato and other vegetables. For precise use of harvested rainwater gravity drip irrigation can be adopted in areas where sufficient head is available to operate the micro tube based drip irrigation system.

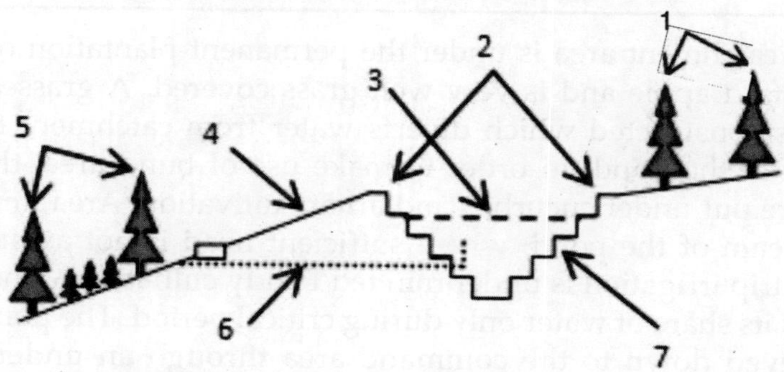

Fig. 3.2 Schematic diagram of water harvesting pond

1. Catchment area (Aonla / Custard apple)
2. Pulses and cucurbits on Bunds
3. Duck and Fish in pond
4. Immediate downstream of pond (Rice/Wheat/Maize)
5. Multi-tier Horticulture (Litchi / mango, guava, vegetables)
6. Take off pipe for Drip Irrigation
7. LDPE film lining covered by stone and soil layer

The harvested rainwater is put to many uses before it is lost to the atmosphere or groundwater. The prominent water users of agriculture may include field crops, horticulture, forestry, fishery, livestock, poultry, duckery, small scale industries besides the domestic needs.

Polythene used in lining of the pond should be UV stabilized and black in color and should have enough strength to resist the weeds. A polythene of 250 to 500 micron thickness can be used to line the pond. Polythene is placed on the staircases of the sides and flat bottom of the pond. Looking at the rainfall pattern of the eastern plateau and hill region conditions a pond of 25 X 20 m size top dimensions with 3.0 m depth is required to harvest runoff from a well maintained catchment area of 2 acre.The total cost of constructing the pond and irrigation system is given below.

Table 3.2 Details of the water harvesting pond

Size	20x25 m
Total depth	3.0 m
Expected depth of water	2.7 m
Volume of water stored	1200 m^3
Lining Material	250-500 µ LDPE
Covering for lining	Soil on stairs

The catchment area is under the permanent plantation of Aonla and Custard apple and is very well grass covered. A grassed water way was constructed which diverts water from catchment towards the inlet of the pond. In order to make use of bund area, the Pond bunds are put under cucurbits and arhar cultivation. Area immediate downstream of the pond, where sufficient head is not available for gravity drip irrigation is under rain fed Paddy cultivation. The paddy plot gets its share of water only during critical period. The pond water is conveyed down to the command area through an underground take off pipe. Further downstream area was planted with litchi based multi-tier cropping system with guava as filler crop. The intercrops like Tomato, Brinjal, Arhar, Groundnut *etc.* were grown.

Table 3.3 System Cost

S.No.	Item	Total cost (Rs. In lac)
1	Cost of pond	1,13,000
2	Cost of drip irrigation system	18,000
3	Planting material	13,000
4	Manures and fertilizers	10,000
5	Plant protection	5,000
	Total system cost	1,59,000

Intercropping of vegetables in winter and other crops in rainy season in juvenile orchard can provide sustainable income to the farmer at early stages. Guava as a filler crop starts yielding from the second year with a rising trend till the main first crop (litchi) comes in to bearing which takes about 7-8 years. With the increase in main fruit (litchi) canopy size the filler crop (guava) plants can be restricted or phased out. The shade tolerant crops like turmeric etc can also be cultivated in the grown up orchard at the later stages. Further, introduction of fishery and duckery in the system meets both the nutritional and economic needs of the farmer. Based on the projections of income in future years the payback period of such a water harvesting system is expected to be less than six years.

The system becomes comparatively more profitable in the areas where land is put to only rain fed cultivation. By adoption of this technology the potential of a land can fully be utilized which enables the farmers to get better income besides fulfilling nutritional requirement.

The catchment area

Parwal and Maize on bunds

Paddy cultivation immediate downstream

Litchi based multi-tier plantation

Fig 3.3 Components of water harvesting system

Farm Ponds

Creation of small water harvesting tanks can significantly contribute to the irrigation potential in hills and plateau region. Major problem associated with water harvesting structures in the region is high seepage. Unlined tanks are prone to heavy seepage losses (300-400 lit per day per m^2) and may not be able to supply water for irrigation during the periods of water stress. Several materials and technologies have been tested and are still being evaluated to check the seepage loss. The cement tank for instance is able to check the seepage losses but the cost becomes the limiting factor. Therefore appropriate lining material which has the twin advantage of checking the seepage and cost effectiveness are required to be promoted in the region. Low density polyethelene (LDPE) film has been used to effectively control the seepage loss at much lower cost than the cement tanks. The film used should have minimum thickness of 200 micron which can sustain physical damage to some extent. To get extended life of plastic film and to avoid the physical damage, the film can be placed on stone pitched surfaces of the pond.

Site selection

The tank construction site should be kept nearer to the command area to avoid excessive conveyance. Due consideration must be given to the location of the available water source that supplies water to the pond. The following points are considered while selecting site for LDPE film lined ponds.

- Location and availability of water source.
- Soil depth: The tank should have minimum depth of 1 meter.
- Present use of the land where tank is proposed.
- Obtainable gravity head
- Higher attainable storage excavation ratio

Excavation of pit

The pits are usually excavated manually. The agricultural implements for soil working such as spade, crowbar and pickaxe are employed in excavation of the pit. The top and bottom dimension are marked on the land using pegs before the excavation process. The excavated soils are kept on side of the tank in embankment form (Fig 3.4).

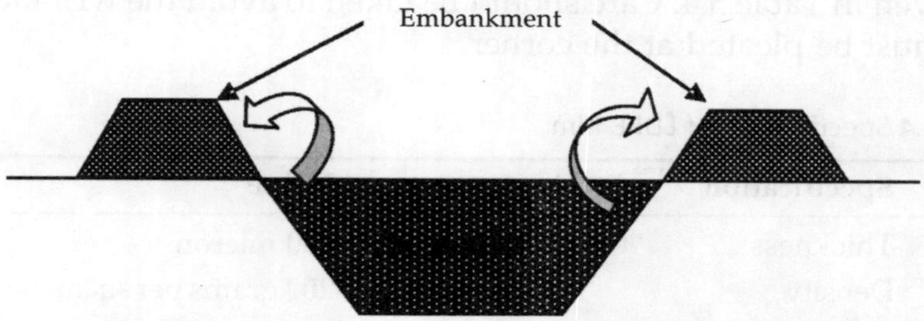

Fig 3.4 Excavation of pit and embankment of the dugout soil

Treatment of pit

Excavation of pit is followed by compaction of bottoms and sides to suppress the angular projections that may puncture the polythene sheet. The side walls and bottom are then treated with 4% atrazine (herbicide solution) so that the plastic film could be saved from puncture caused by root infestation. After these, the surfaces of side wall and bottom are smoothened using dung and soil mixed in water.

Excavated pits or small ponds are easily constructed in the relatively flat areas with deep soils. Though the soil compaction reduces the seepage through the wall of the tanks, but it is not sufficient to retain the water for longer time. Therefore, a water barrier must be used to further control the seepage. The lining of plastic film (LDPE) ensures the prevention of the seepage of the stored water. Thus, water can be stored for longer time. The tank should be excavated in such a way that it gives the maximum storage to excavation ratio in order to minimize the storage cost of water. It can be achieved when the excavated material are put over the bank of the pit to raise the wall of the pit. The tank should also be designed for minimum evaporation loss. The evaporation loss can be minimized, when the ratio between the exposed water surface and storage volume is minimum. Thus, evaporation loss can be reduced by making the wall of the excavated pit vertical, but it may not be stable. The side slope should be optimized in order to minimize the evaporation loss from the tank. The complete vertical tank can be obtained using cement concrete and brick work, but has the limitation of higher initial cost.

Laying of LDPE film

The LDPE (Low Density Poly Ethylene) films are spread over the excavated pit (Fig 3.5b). The minimum specifications of the LDPE film are given in Table 3.4. Care should be taken to avoid the wrinkles. The film must be pleated at the corner.

Table 3.4 Specification of LDPE film

S.No.	Specification	Value
1	Thickness	200 micron
2	Density	200 grams per sq. meter
3	Colour	Preferably black

Jointing of film

The LDPE films are joined using bitumen (normally used in hot mix road construction). To jointing, both films are spread on the flat surface. The corner 15 cm along the length of the both film then scrubbed using emery paper or sand paper (120 grade). The bitumen heated to 80°C when it started flowing freely. This hot bitumen then pasted over the scrubbed corner of the film and overlapped (15 cm) both the film and some weight are kept over for setting. After 30 minutes, the film joins firmly. The LDPE films can be joined with the help of electric press also.

Pond Outlet

The film punctured at the bottom corner and 25 to 40 mm pipe (GI, PVC or HDPE) inserted inside the tank from tank side. To fix the pipe with the film, 10 cm thick CC is applied. Outside of the tank one gate valve is provided to regulate the flow from the tank.

Anchoring the film

The LDPE film has the lower dart impact resistance (110 gram; ASTM D1709) and thus prone to puncture. To avoid the puncture, it is necessary to safeguard the film. The method of stone pitching is extensively used in anchoring the film. However, other material such as tarfelt sheet (mica impregnated coal tar sheet) can also be used for covering of film.

Stone pitching

The stone pitching involve the arrangement of stones (10 to 20 cm boulder with no conical projections at the side of the tank. These boulders are available in the river bed of the hilly terrain. The stone pitching is recommended where the side slope of the tank is higher than 1:1. The larger boulders are arranged at the lower side and gradually smaller boulders are arranged towards the top. Cement pointing is provided to fix the boulder and 10 cm thick soil layer is spread on the bottom.

(a) Excavation of pit (b) Lining with LDPE film (c) LDPE lined tank filled with water

Fig 3.5 Construction procedure for farm ponds

Plastic Lined Doba

The area under the fruit orchards is increasing at faster rate in the eastern plateau region. To make the profitable use of less productive uplands, the farmers of eastern plateau and hill region have recently shifted their focus from paddy-pulse cropping sequence to cultivation of fruit based farming systems. Generally the saplings are planted during the months of July to September. Initially there is no dearth of water and the plants grow well with the available monsoon water. As the monsoon withdraws, water drains out of the area and soils of midlands and uplands start losing moisture. In the months of April to May plants face severe water stress and unless some watering is done it is practically impossible for plant to survive. Although, use of soil moisture conservation techniques for ground water recharge are efficient tools for improving plant growth of fruit trees of more than 2 years of age, the raised water level is not enough for supplying moisture to shallow roots of newly planted trees. To combat such water stress to the newly established orchards, a plastic lined *Doba* technology has been developed at ICAR-RCER. This is low cost technology, easy to understand, construct and maintain.

What is Plastic lined *Doba*?

Plastic lined *Doba* is a polythene lined (200 micron UV-stabilized black polythene) small pond for harvesting of direct rainfall. The structure can store 4500 litre of water. The tank is lined with black polythene to check the loss of water due to seepage. The structure is constructed before the onset of monsoon for collection of rainwater. To avoid the silting of *Doba* due to runoff water a small bund is formed along the border to stop the entry of surface runoff in it to the tank. The average annual rainfall of the region is about 1400 mm, and the depth of *Doba* is 1000 mm. Therefore there is every possibility that the structure will get water to its full capacity in one monsoon season in spite of evaporation loss of 1.5-2.0mm/day in rainy season. The dimension of *Doba* has been designed in such a way that, it can provide lifesaving irrigation to 10 plants for six months. The daily average requirement of one mango plant for life saving irrigation is 1 litre and for six months 1800 litre of water is required by 10 number of mango plants. The raised bunds of *Doba* can be utilized for growing cucurbitaceous crop with the residual moisture.

Specifications

Size of the *Doba* is 3.0 m length 1.5 m width and 1.0 m depth. The *Doba* is lined with black polythene of 200 micron thickness. Polythene of 6 m length and 4.5 m width is required to line the structure. The structure is covered from above with thatch (roof) of size 3.5 m length and 2 m width. A small opening is kept in the thatch to take out the water for irrigation.

Construction

Doba is constructed with simple tools available with the farmers, mainly a spade and bucket. Before going for actual digging of the tanks for *Doba,* a proper layout of the field is made in such a way that one *Doba* covers nearest ten plants. The spacing of the *Doba*s in orchard is made accordingly. While preparing a layout, factors like movement of machinery, ease of access to *Doba* should also be kept in mind. After finalizing the locations, the layout of the structure is marked on the ground with the help of a measuring tape and ropes.

Digging of pit: The loose soil from the digging of *Doba* can be dumped around the structure to form a small bund of height of about 15-20 cm. This bund should be at least 45 to 60 cm away from the structure. After completion of digging, the sides and the bottom are smoothened. Any pointed material that can puncture the polythene should be removed. If the soil is very rough, mud or slurry prepared from mixture of soil and water can be applied on sides and bottom for making the surface smooth.

Placing the polythene in structure: A piece of size 6 x 4.5 m of 200 micron black polythene is cut and placed in to the *Doba* structure. There may be formation of wrinkles at the corners; therefore proper folding of polythene is to be done at corners to avoid excessive wrinkles. For proper anchorage, the polythene is placed in the trench dug on the border of the *Doba* and soil is filled above it. This makes sure that the polythene doses not slide down due to weight of water.

Fixing the polythene in trench: A trench of size 20 cm wide and 15 cm deep is dug around the border of the *Doba* after leaving a margin of 20-30 cm. This trench is used for fixing the polythene at the top to avoid sliding. The extra polythene remaining above the depth of structure is placed firmly in this trench and soil is filled above it. The weight of the soil does not allow the polythene to slide.

Digging of trench for fixing polythene | Fixing the polythene in Doba | Fixing of polythene in trench

Fig. 3.6 Procedure of Doba construction

Covering with thatch: The *Doba* is covered from above with a thatch prepared from locally available material like straw and bamboo. The paddy straw or dry grass is placed between two bamboo frames of size 4x2.5 m size. A small opening of size about 50 cm x 50 cm is kept in the thatch to take out water for irrigation. The thatch avoids entry of foreign matter in the stored water as well it reduces the loss of water by evaporation. Thatch also prevents the risk of animals falling in to the *Doba*.

Benefits of Plastic lined Doba technology

- *Water Conservation:* The first and foremost benefit is the conservation of water. It collects direct rainfall volume of about 4500 lit in one monsoon season. Since the structure is polythene lined and is covered with a thatch the loss of water through seepage and evaporation is also minimal.
- *Lifesaving irrigation:* The conserved water during the monsoon season can be used to provide the lifesaving irrigation to fruit plants during dry periods.
- *Reduced mortality:* The provision of life saving irrigation greatly reduces the plant mortality due to reduced water stress in the root zone of the plant.
- *Good crop health:* Due to assured irrigation during the dry periods the plant maintains good health which imparts disease and pest resistance to the plant.

Problems associated with Plastic lined Doba

There can be a problem of damage to the polythene or small puncture due to rat or any other rodents. This can be repaired easily using the simple procedure. Removing the whole sheet of plastic from the structure; take a piece of plastic sheet from the side of the existing plastic and place it over affected area. These two can be joined using some heating object like electric press. The hot press makes the polythene to melt partially and they get joined forming a leak proof joint. Further the damage caused by termite and ants in the polythene can be controlled by spraying Biflex @ 2ml/liter of water on the walls and bottom of *doba* at the time of construction and fixing of polythene.

Assessment of the technology in farmers' field

A field survey of water levels in Doba during the month of May under farmers' field showed that only 15% *Doba* were dry, 45 % *Doba* were having water levels between 5 to 20 cm (*i.e.* 225 to 900 lit of water), 25% *Doba* were having water levels between 21 to 40 cm (945 to 1800 lit of water) and 15% *Doba* were having water levels between 40 to 50cm (1800 to 2250 lit of water). Eighty percent of the structures retained water to depths ranging from 5 to 45 cm (225 lit to 2025 lit) by the end of the May which shows the availability of water throughout the water stress period.

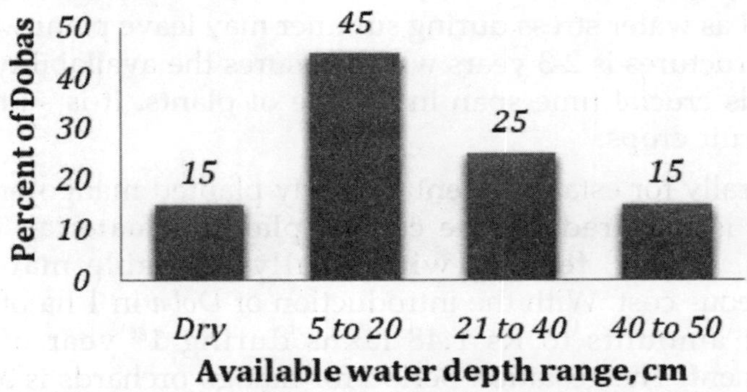

Fig. 3.7 Water level in *Doba* at the end of May-2009

Mango plants in the orchard having 5 *Doba*/acre for 60 numbers of plants showed better plant height, girth at graft union (GGU), canopy diameter and per cent survival of plants compared to that with 2 *Doba*/acre and no *Doba*.

Fig. 3.8 Plant status with and without Doba

Economics of *Doba* technology

A total of two mandays are required for digging of *Doba* which costs about Rs.300/- and the cost of polythene is Rs. 780/- (6 kg @ Rs.130 per kg) and thatch (roof) may be prepared from farm waste with minimum additional cost of about Rs.120/- for paddy straw and bamboo. So the total cost of one such *Doba* comes out to be Rs.1200/. If farmer is having an hectare area under fruits planted at a spacing of 5 x 5 m (400 plants per ha), the required number of *Doba* to meet the irrigation requirement will be 40 and the total cost for one hectare will be Rs.48,000/-. Water saving through *Doba* technology has been worked out to be 1,80,000 liter/ha. Initial three years of fruit plants are crucial as water stress during summer may leave plant wilted. Life of *Doba* structures is 2-3 years which ensures the availability of water during this crucial time span in the life of plants. It is suited for all orchard fruit crops.

Generally for establishment of newly planted mango orchard, Rs 1.00 lakh is required for the cost of planting material, fertilizer, pesticide, labour, fencing with locally available material and miscellaneous cost. With the introduction of *Doba* in 1 ha of land, the total cost amounts to Rs 1.48 lakhs during 1st year of orchard establishment. The gestation period for mango orchards is 3 year and with *Doba* technology, actual benefit will start to accrue after 4 years at the estimated rate of Rs 1.00 lakh per hectare (Yield: 4000 kg/ha) from mango orchard. The inter space between mango plants are utilized for growing intercrops during the initial period and provides some remuneration to overcome cost incurred in the establishment and management of orchard. Further the harvested residue of intercrop is used as dry mulch in the surface of root zone of mango plant to conserve water. The payback period for mango orchard is 4 year i.e. the cost involved for the establishment is met from the income at the 5th year of mango orchard.

Table 3.4 Cost of cultivation of mango/ha using *Doba* technology

Particular	Quantity	Unit	Cost (Rs)
Capital cost of *Doba*/ha	40 nos.	@ Rs 1200	48,000
Cost of planting material + fertilizer + pesticide + labour cost + miscellaneous cost for mango orchard	1 ha	@ 1 lakh/ha	1,00,000
	Total cost		**Rs 1,48,000.00**

Owing to its low cost and ease of construction, the *Doba* (Jalkunda) technology was readily accepted by the farmers. Farmers have agreed to convert their low productive land into mango orchards and use these *Doba* to harvest rainwater during establishment phase of the orchards. Lifesaving irrigations from *Doba* has resulted in no mortality of plants particularly during the summer months and better growth. The plastic lined *Doba* are also being used by farmers for the purpose of storing water from seasonal streams and using the water after drying up of water from the streams during the summer season. This technology taken to the farmers has helped them to increase their income and also to diversify the crops to meet their nutritional requirements.

4

Irrigation Management in Fruit Crops

Water is essential for human civilization, living organisms, and natural habitat. Agro ecosystems, which are the main consumers of fresh water in India, currently face the challenge of developing new approaches to water management that will insure the protection of water resources and their integrity for their future use. Fruit crops, which have gained commercial tone in the recent years, are important component of Indian agriculture. Due to low irrigation and water use efficiencies, irrigation managers are looking for new and effective approaches to fruit crop water management. The studies suggest that it is possible to increase the water use efficiency even at existing level of overall irrigation efficiency. Irrigation of agricultural lands accounts for over 85% of water usage worldwide. A minor reduction in irrigation water use or slight improvement in efficiency could substantially increase the water savings simultaneously improving water availability for other purposes. Optimal irrigation is of critical importance for agriculture with decreasing freshwater resources in times of climate change. Despite of huge water allocation for irrigation, water productivity in the agricultural sector is very poor with water losses accounting to more than 50% total applied water. So new adaptations in irrigation strategies is necessary to improve the performance of agricultural water use.

Management of irrigation is an important practice in fruit production regions of the world, particularly in arid and semi-arid climates. It draws attention to the principles of irrigating fruit crops, and identifies the critical factors for measuring crop requirements and the control of irrigation, both to optimize cropping and minimize waste. Frequency of irrigation and amount of water applied to fruit crops is a common concern among the fruit producers of India. Insufficient water supply may result in reduced tree growth, yield and fruit quality due to water stress. Excessive irrigation, on the other hand, may increase nutrient leaching, water-logging problems, incidence of pests and diseases and the associated cost of frequent operation and maintenance of the irrigation system. Optimizing water applications by scheduling irrigation to fruit orchards may increase water productivity, reduce production costs, and increase tree growth and fruit yield. Irrigation plays an important role in raising and stabilizing yield and maintaining quality of fruit crops.

The Concept of Deficit Irrigation (DI)

Traditionally, agricultural research has focused primarily on maximizing total production. In recent years, focus has shifted to the limiting factors in production systems, notably the availability of either land or water. Within this context, deficit irrigation (DI) has been widely investigated as a valuable strategy. Deficit irrigation is an optimization strategy in which irrigation is applied during drought-sensitive growth stages of a crop. Outside these periods, irrigation is limited or even unnecessary if rainfall provides a minimum supply of water. Water restriction is limited to drought-tolerant phenological stages, often the vegetative stages and the late ripening period. Total irrigation application is therefore not proportional to irrigation requirements throughout the crop cycle. While this inevitably results in plant drought stress and consequently in production loss, DI maximizes water productivity, which is the main limiting factor (English, 1990). In other words, DI aims at stabilizing yields and at obtaining maximum WP rather than maximum yields

The deficit irrigation can be obtained (i) by increasing the interval between two irrigations, (ii) by reducing the amount of water applied during each of the irrigation or (iii) by combination of these two strategies. While scheduling the deficit irrigation for a particular crop, the stress should not become so severe that the plant suffers permanently; instead, it should be moderate beneficial stress to which plant is subjected during non-critical phenological stages. To avoid the severe water stress, accurate assessment of plant water stress is essential. As mentioned earlier plant is the best indicator of its water stress and plant based indicators can be used for deficit irrigation scheduling.

Partial Rootzone Drying Irrigation

In recent years, deficit irrigation is seen as one of the best alternatives for crop production under water scarcity conditions. Water deficits will induce changes in the gas exchange characteristics of a plant, particularly it will affect the stomata resulting in reduced the evaporative demand of plants. This effect of water stress (deficit) would obviously be an advantage if it could be achieved without any detrimental impact on the crop performance. Regulated deficit irrigation and partial root zone drying are the two techniques subjecting the plants to deliberate water stress such that water requirement of the plants is reduced without much impact on the crop yields.

Many investigations have been conducted to assess the impact of deficit irrigation on crop performances, efficiency of water use and profitability. This aspects is further researched with varying crops and under varying crops. Full irrigation (FI) is scheduled to meet the full evapotranspiration requirements of crop that result in maximum yield. Now-a-days, full irrigation is considered a luxury use of water that can be reduced with minor or no effect on profitable yield (Kang and Zhang, 2004). Deficit irrigation is a water saving technique used in irrigated agriculture to improve the water productivity (WP). Deficit irrigation (DI) and partial root-zone drying irrigation (PRD) are the water-saving irrigation methods that cut down irrigation amounts of full irrigation to crops. The amounts of irrigation reduction is crop-dependent and generally accompanied by no or minor yield loss that increases the water productivity (Ahmadi *et al.*, 2010b).

The Concept of PRD

Partial rootzone drying is the practice of using irrigation to alternately wet and dry two spatially explicit parts of the plant root system. It creates simultaneous wet and dry (or drying) areas within the root zone. Irrigation is applied on one side of the plant row and allowing the other side to dry out keeping only one side (zone) moist at any one time. The irrigations are then alternated to the dry side after a set period of time and the cycle of altering the wet and dry sides is continued for the set time periods.

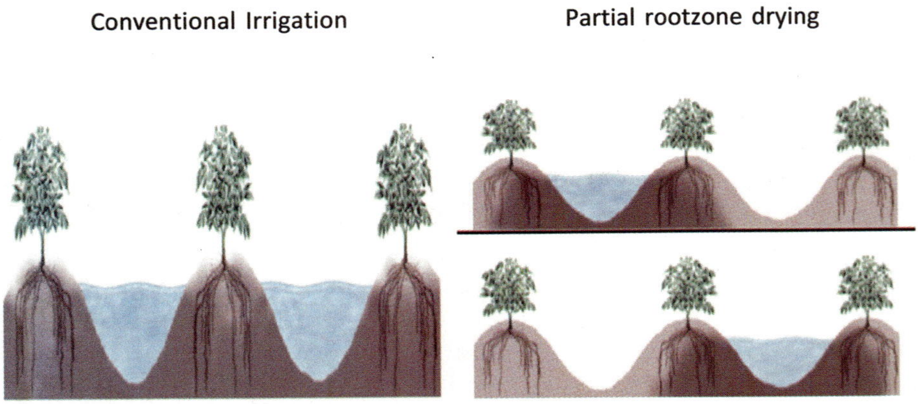

Fig 4.1 Water application under conventional and PRD technologies

Under this type of wetting pattern chemical signals produced by the roots in the dry zone will reduce stomatal aperture in the plant leaf. At the same time the fully hydrated roots maintain a favorable water status throughout the aerial parts of the plant. The re-wetting frequency under PRD should be determined by detection or prediction of completion in extraction of soil water from the drying side. In practice this can be identified from soil moisture depletion or estimates of plant water use based on evaporative conditions or measurement of sap flow. PRD uses biochemical responses of plants to water stress to achieve a balance between vegetative and reproductive development.

Crop physiology under PRD

Under water stress there are certain subtle changes in the physiology of the crop that leads affects the crop metabolism. This change is mainly induced in terms of variation in the abscisic acid production, photosynthesis and stomatal conductance.

Photosynthesis and stomatal conductance

This technique is based on two principal assumptions. First is that fully irrigated plants have widely opened stomata such that a small narrowing of the stomatal opening may reduce water loss substantially with little effect on photosynthesis (Jones, 1992). The other assumption is that the plant roots in the drying side of the rootzone can respond to drying by sending a signal to shoots where stomata may be closed to reduce water loss (Davies and Zhang, 1991). With this type of mixing signals it is possible to separate the biochemical and physical effects of reduced water availability. In addition to reduced stomatal conductance, it is noted that shoot extension is also inhibited, increasing the yield to biomass ratio. Water is lost as transpiration and CO_2 is absorbed for photosynthesis through stomata. Therefore, any variation in stomata opening affects stomatal conductance (gs) and photosynthesis rate (An). Reduced stomatal conductance in early stages of water stress inhibits transpiration rate more than it reduces the intercellular CO_2 concentration that is the driving factor for photosynthesis.

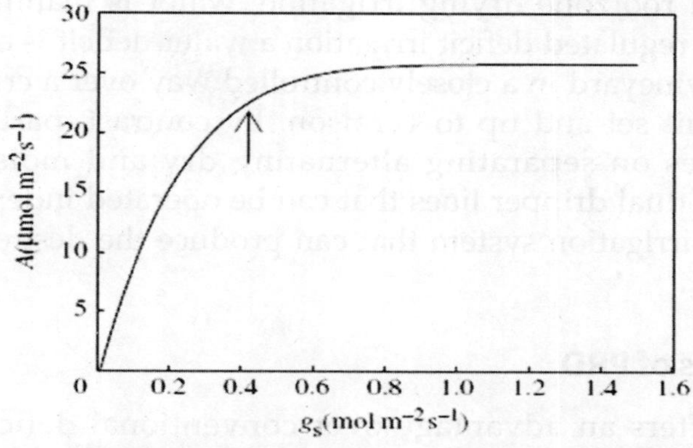

Fig 4.2 Relationship between net CO_2 assimilation rate (A) and stomatal conductance (g_s)

Abscisic acid production

Research into the physiological changes that occur during water stress has led to improved understanding of plant response to stress in terms of chemical signals passing from roots to leaves. The plants first line of defense when faced with water shortage is to close its stomata to conserve moisture. One of the principal compounds that elicits this response is abscisic acid. As soil water availability reduces, this acid is synthesized in the drying roots and transported to the leaves in the transpiration stream. Abscisic acid derived from drying roots is thought to be responsible for the physiological changes to the plant as well as the inhibition of the plant hormones cytokinins that affect stomatal aperture and lateral shoot development. Stomata respond by reducing aperture, thereby restricting water loss. Improvement in WUE results from partial stomatal closure. However, an inevitable consequence is reduced photosynthesis, as carbon dioxide and water vapour share the stomatal pathway through the leaf surface.

How PRD different from RDI

Regulated deficit irrigation and partial rootzone drying are designed to limit vegetative vigour and improve water use efficiency in perennial crop plants such as grapevines and fruit trees. This shift in crop management, compared with traditional irrigation methods which use more water, is a feature common to both regulated deficit irrigation and partial rootzone drying. These two methods of irrigation do, however, differ fundamentally in two key respects. With regulated deficit irrigation water application is manipulated over time whereas, with partial rootzone drying irrigation, water is manipulated over space. With regulated deficit irrigation a water deficit is applied in an orchard or vineyard in a closely controlled way over a critical period, i.e. after fruit set and up to veraison. By contrast, partial rootzone drying relies on separating alternating dry and moist roots and necessitates dual dripper lines that can be operated independently or some other irrigation system that can produce the desired pattern of soil wetting.

Advantages of PRD

PRD offers an advantage over conventional deficit irrigation strategies because it helps reduce water use by separating the biochemical responses to water deficit from the physical effects of reduced water availability, allowing developmental processes associated with plant growth to remain unperturbed. Irrigating a reduced root zone volume in this way reduces crop water requirement. Further the PRD leads to preferential root growth. Proper aeration in the crop root zone enhances the nutrient uptake in from the drying zone. PRD increases root hydraulic conductivity by cell-to-cell channels known as aquaporins. Previous research has confirmed that approximately up to 30-50% of water saving can be achieved under PRD with no significant yield reduction. Reports showed that PRD advanced 5–7 days ahead of scheduled harvest leading to better price for the farm produce.

Limitations of PRD

It is important to understand that this technique pushes the crop to its limits and should only be applied to well-established healthy trees. Looking into technical aspects, the cost of implementing PRD irrigation systems is comparatively higher than traditional systems. It requires continuous monitoring of soil moisture in the root zone and high-level management skills are required in scheduling the irrigations and managing the water application patterns at field level. Faulty management of the system may lead to considerable yield and economic loss to the farmer.

Irrigation practices under PRD

Application of irrigation water under PRD technique requires provision of furrow or drip laterals on both sides of the plant rows so that the alternate wetting and drying can be achieved by altering the irrigation events. Imposing the cycle of PRD can be targeted to a particular tree growth phase but would usually be maintained during an entire growing season. The duration of the wetting and drying cycles should not be too short (less than 3 days) nor too long (more than about 15 days). It is a common practice to have the PRD cycle of 10 to 15 days (Davies *et al.*, 2000; Stoll *et al.*, 2000). Switching of the sides of irrigation should be based on threshold soil water content at which maximum xylem abscisic acid concentration is produced. This needs careful understanding of the Abscisic acid production levels under different soil moisture levels.

Fig 4.3 Alternating irrigation among the furrows to get the effect of PRD

Yield and quality improvement

The expectation from PRD is that any yield reduction will be insigni?cant compared with the benefits gained through diverting the saved water to irrigate other crops. If deficit irrigation is properly practiced, it may increase crop quality. PRD has been found to be promising in several crops. For example, the protein content and baking quality of wheat, the length and strength of cotton ?bers, and the sucrose concentration of sugar beet and grape all increase under deficit irrigation. According to Kang *et al.* (1998) when the two halves of a maize root system were alternately exposed to drying and wetting, water use was reduced by 34.4–36.8% and total biomass production was reduced by only 6–11%, as compared with well-watered plants.

Performance of table grapes was assessed under conventional drip irrigation (CDI), irrigated on both sides of the root system, alternate partial root-zone drip irrigation (ADI), irrigated with 50% of the control, alternatively on the two sides of the root system during consecutive watering and fixed partial root-zone drip irrigation (FDI), irrigated with 50% of the control on only one side of the root system(Du *et al.*, 2008). Fruit yield and yield components under different irrigation treatments in 2005 and 2006 are shown in Table 4.1.ADI gained 44.8% more bunch number per tree than FDI in 2005 and it also improved bunch number per tree significantly as compared to CDI in 2006. Berry number per bunch was at the same level under the three irrigation treatments in 2005 but ADI gained 30.7% more berry number per bunch than FDI at the same irrigation level in 2006. Bunch weight and volume of ADI was kept at the same level with CDI in 2005 but it was reduced significantly in 2006. Furthermore, ADI increased the percentage of edible grape by 3.88–5.78% significantly at the same fruit yield level in both years.

Table 4.1 Yield attributes of table grape underdifferent irrigation methods.

Parameters	Year	ADI	CDI	FDI
Bunch number per tree	2005	5.89 a	5.75 ab	3.25 c
	2006	25.30 a	19.75 b	23.50 ab
Berry No. per bunch	2005	102.30 a	116.00 a	112.00 a
	2006	66.08 a	57.80 ab	50.56 b
Bunch weight (g)	2005	486.80 a	519.47 a	442.69 a
	2006	355.82 b	445.83 a	360.22 b
Bunch volume (cm3)	2005	427.16 a	473.69 a	362.41 a
	2006	339.88 b	415.82 a	323.23 b
Berry weight (g)	2005	4.07 a	4.08 a	3.95 b
	2006	5.38 b	7.71 a	7.13 a

% of edible grape	2005	93.36 a	87.58 b	95.47 a
	2006	94.25 a	90.37 b	93.23 a
Yield per tree (kg)	2005	5.18a	5.06 ab	2.66 c
	2006	17.30 a	17.08 ab	16.25 b
WUEET (kg/m3)	2005	3.28a	2.24	1.69c
	2006	7.11a	5.61	5.61

Under ADI, the vitamin C (Vc) content of fruit is increased by 15.3–42.2% in both years, but decreased the fruit titratable acid (TA) by 13.2% at the same level of TSSC compared to CDI, therefore increased TSSC/TA by 22.5% at post-veraison in 2005. ADI also increased TSSC at the similar level of TA with highest TSSC/TA significantly at harvest in 2006.During both the year higher Vc was observed under ADI indicating the, indicating a higher nutrition value in the berries produced under ADI. Higher TSSC/TA ratio under ADI indicates higher total soluble solids concentration and medium titrated acid of grape, indicating sweeter and more delicious fruit in both years.

Table 4.2 Fruit quality of table grape under different irrigation regimes

Year	Treatments	Vc (mg/ 100 FWg)	TSSC (%)	TA (mg/ 100 FWg)	TSSC/ TA
2005	CDI	0.268 b	13.56 b	5.16 a	26.28 c
	ADI	0.309 a	13.68 b	4.48 b	32.19 b
	FDI	0.162 c	15.00 a	4.25 c	33.48 a
2006	CDI	0.090 b	13.17 b	2.33 a	56.52 b
	ADI	0.128 a	13.65 a	2.33 a	58.58 a
	FDI	0.108 ab	13.10 b	2.47 a	53.04 c

Table 4.3 Fruit quality attributes of apple cv. Fuji under different irrigation regimes

Days after harvest	Treatments	Firmness (N)	Soluble solids (Brix)
1	Control irrigation (CI)	69.1	13.50
	Deficit irrigation (DI)	73.2	15.08
	PRD	74.3	15.55
	LSD (a=0.05)	3.3	0.75
7	Control irrigation (CI)	73.3	14.40
	Deficit irrigation (DI)	75.0	15.75
	PRD	75.4	16.60
	LSD (a=0.05)	NS	0.68

Performance of apple was assessed under three irrigation treatments, conventional irrigation (CI) irrigated on both sides of the tree; a partial root-zone drying treatment (PRD) which received 50% of CI on alternating sides of the tree and a deficit treatment (DI) which received 50% of conventional irrigation on a fixed (east) side of the tree (Leib *et al.*, 2006). Compared to the CI regime, reduced fruit weight and higher total soluble solids was observed in the fruits under PRD. No significant difference in flesh firmness or starch score due to irrigation treatment was evident in either season. The development of red fruit skin colour, assessed by the percentage of green skin was similar between irrigation regimes and seasons, although fruit in Year 2 were rated as having greater intensity of redness.

Fruit quality of mango also showed significant improvement over conventional and regulated deficit irrigation regimes. Application of PRD at 50 % of the full crop water requirement (ETc) resulted in higher fraction of fruit (Spreer *et al.*, 2007). Differences were consistently significant as compared to the control. When mango fruits are produced for processing, a higher fraction of mesocarp is advantageous-provided that differences prove to be consistent over an extended period.

Table 4.4 Effect of irrigation on mesocarp volume of mango fruit

Treatments	2004	2005
CO (100% of ETc)	71.1 a	76.1 a
PRD (50% of ETc)	78.0 b	78.5 b
RDI (50% of ETc)	77.3 b	76.3 a
NI (no irrigation)	71.1 a	76.3 a

It is well proved that PRD has the potential to be used in arid regions with limited water sources. PRD is a promising water management strategy for saving water in drought-prone regions. It allows enhanced water use efficiency in crop production by exploiting the plant's long-distance signaling mechanisms that modify plant growth, development and functioning as the soil dries. Improved water use efficiency leads to improved water productivity.This will mean that all users will need to increase water and irrigation management efficiency to make the most of this basic resource. Environmental considerations are increasingly playing a part in the growing and marketing of crops. Careful and effective crop monitoring and irrigation control will be critical to profitable cropping in the future.

Trunk Diameter Based Irrigation Scheduling

The success of deficit irrigation depends on accurate assessment of plant's water stress. Among the strategies that can be applied to attain this objective is, first and foremost, the saving of water through the improved determination of crop water requirements and the development of biological and physical criteria that lead to improved, innovative and precise irrigation management practices. Plant itself is a better indicator of its water stress (Fernandez and Cuevas, 2010) and plant based water stress indicators can be used for scheduling deficit irrigation. Earlier studies assessed the effect of water deficit on trunk diameter variation (TDV) and identified the most sensitive index derived from TDV (Ortuno *et al.*, 2006). Some recent studies have demonstrated the applicability of trunk diameter variation (TDV) based indicators to detect the plant water stress.

Deficit irrigation scheduling based on soil water measurements requires good knowledge of root distribution and relative density, with the limitation that uncertainty increases when the wetted soil volume varies in three dimensions, as it does in trickle irrigation. For irrigation scheduling procedures involving high-frequency irrigation systems, information about crop evapotranspiration (ETc) is required. Here, too, uncertainties abound due to the effects of tree canopy architecture, the degree of soil canopy cover and soil surface management. Also, the estimation of ETc in young tree plantations, where it is necessary to use empirical factors or reduction coefficients or where there are actively growing weeds, involves uncertainty when irrigation scheduling is based on this parameter. For these reasons, the use of plant-based water status indicators has become very popular in recent years for studying plant–water relations and for planning more precise irrigation programs, because it is recognized that the tree itself is the best indicator of its water status. Since the plant water status controls many physiological processes and crop productivity, this information can be highly useful in irrigation scheduling (Goldhamer and Fereres, 2001; Goldhamer *et al.*, 2003). Particularly under deficit irrigation conditions, the continuous control of tree water status is crucial in order to prevent a moderate, potentially beneficial, water stress from becoming too severe and ending in a reduction of the yield reduction.

Plant Based Indicators of water stress

A selected list of direct and indirect plant indicators of water stress is shown in Table 4.5. How well an indicator responds to accumulated water deficit in soil and/or atmosphere is shown as sensitivity. The classical physiological measurements of leaf water potential (with a pressure chamber on detached leaves) and stomatal conductance (with a diffusive porometer or chamber methods for intact leaves) are only moderately sensitive even if these are based on theoretically sound approaches and have good precision. Similarly, stomatal conductance can become stable when the plants are able to adapt to water deficits (Moriana and Fereres, 2002). Lack of sensitivity of some of these measures with soil moisture and high labour costs have restricted their widespread use for commercial irrigation scheduling..

Table 4.5 Direct and indirect plant based indicators of water stress

Indicator	Sensitivity	Commercial use
Direct – measures plant water status		
Leaf water potential	Moderate	Poor
Stem water potential	High	Poor
Sap flow measurement	High	Good
Indirect – measures plant response to PWS		
Canopy Temperature	Moderate	Good
Stomatal Conductance	Moderate	Poor
Trunk Diameter	High	Good

Stem water potential measured on detached, bagged and equilibrated leaves has emerged as a more robust indicator of plant water status (Chone *et al.*, 2001) because it is an integrator of the whole plant transpiration, soil water availability, soil hydraulic conductivity and plant capacity to transport water from the soil to the atmosphere. However, it suffers from the same limitation as leaf water potential in its applicability to commercial irrigation scheduling. Continuous measurement of sap flow rates in plant stems can be a good indicator of water deficit in plants as sap flow can be quite sensitive to stomatal conductance, transpiration and atmospheric demand (Misra and Sands, 1992). It is also feasible to design irrigation scheduling schemes using controlled threshold values of sap flow rates to detect onset of water stress for specific crops. Growth rate measurements of either the whole or part of the plant are considered to be very sensitive indicators of water stress, although measurements can be difficult and interpretation needs experience. With the recent advance in linear variable displacement transducers, it is possible to measure growth rates of stems and fruits from diameter measurements (Moriana and Fereres, 2002). Such measurements is expected to help develop control systems for irrigation using critical values of growth rates for well irrigated and water stressed plants.

Basics of Trunk Diameter Variation (TDV)

When transpiration (Ep) begins early in the morning, a tension is created in the xylem from the evaporative surface of the leaves to every organ of the plant. Part of the water stored in the plant tissues during the night is then lost, allowing the plant to respond rapidly to changes in atmospheric demand, without the need to rely on water uptake by the roots, which starts later (Hinckley and Bruckerhoff, 1975). This affects every water-storing organ of the plant, so diurnal diameter changes occur in all parts of the plant, including the stem, branches, roots, leaves, and fruits. For large plants such as trees, the water stored within the trunk may contribute substantially to their Ep. Water from the phloem and related tissues (cambium and green tissues of the bark, made up predominantly of parenchyma cells), as well as from the living tissues of the outer xylem (Fig. 4.4), is withdrawn and lost by Ep, so that the trunk diameter decreases. Changes in the water content of extensible tissues of the stem are readily reversible, TDV. Such variations are driven by changing water potential in xylem. Water can also be withdrawn from the inner, woody tissues of the xylem, but then cavitation rather than shrinkage occurs, as this tissue is less elastic.

The work by Herzog *et al.* (1995) illustrates the diurnal dynamics of SDV, together with those of the sap flow (SF) through the stem and the simultaneous exchange of water between the xylem and the phloem and related tissues. They reported five distinct phases in the diurnal courses of SF and SDV, as shown in Fig. 4.5A. Phase I describes the stem's rehydration at night, when the lowest SF rates are recorded. The internal stores replenish, and the stem radius increases. Phase II describes the lag between the increase in flow soon after dawn and the shrinking of the stem. Phase III is the period during which the flow through the stem increases to its daily maximum and the stem shrinks rapidly. Phase IV describes the lag between the maximum in the flow and the minimum in the radius at its reversal point in the afternoon. Finally, Phase V is characterized by a decrease in the flow and an increase in the stem radius. This period may last for a few hours after sunset, if the tree has transpired a lot during the day.

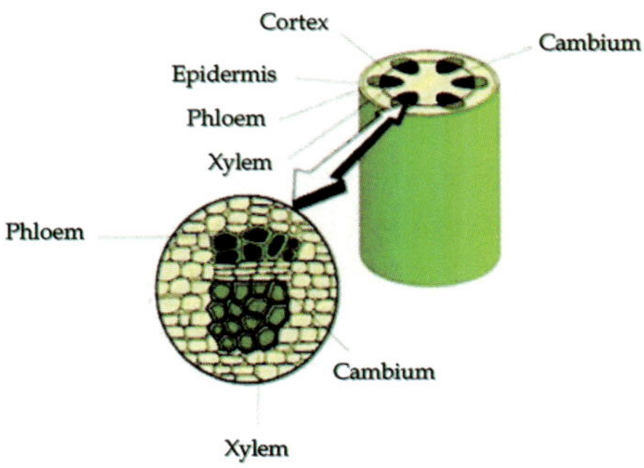

Fig. 4.4 Cross section of the tree trunk

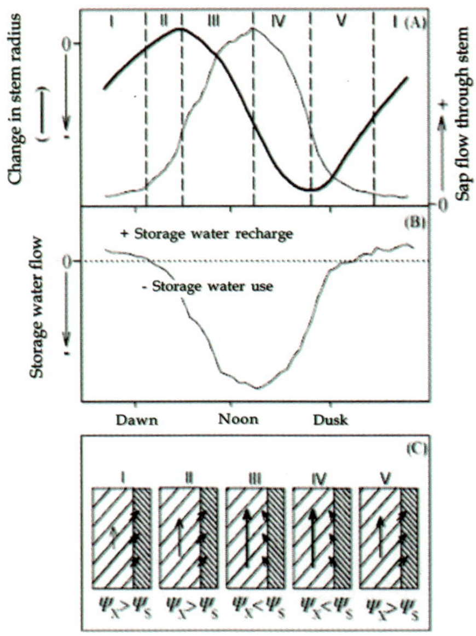

Fig. 4.5 Daily dynamics of sap flow and stem radius.

Fig. 4.5B shows the expected exchange of water between the xylem and the phloem and related storage tissues for each of the five phases described in Fig. 4.5C. During Phases III and IV, water is withdrawn from the phloem and related tissues, due to a decrease in $_x$; consequently, the stem radius decreases. Water is withdrawn from the storage tissues most readily during Phase III, after the nocturnal rehydration in Phases I and II. During Phase IV, withdrawal of water becomes progressively restrained. At the beginning of Phase V, the exchange of water changes direction, due to $_x$ becoming greater than the water potential in the phloem and related tissues. The nocturnal rehydration starts in this phase, and continues throughout Phases I and II; simultaneously, the stem diameter increases.

TDV Sensing Devices

The first stem growth readings for scientific purposes were low-frequency, low-precision records using tapes or calipers. The first highly sensitive dendrometers and dendrographs were developed in the late 1800s but the technique did not become popular until it was simplified by modern technology. Radial changes were monitored with a linear motion potentiometer, and with a variable differential transformer. Radial dendrometers are ideal for monitoring the activity of smaller portions of the cambial cylinder, while diametric and circumferential devices provide a larger sample that can be more indicative of total stem growth. Recording SDV for assessing water stress in plants increased considerably when technology enabled precise measurements, originating the so-called micro-morphometric method, defined as the accurate measurement and continuous recording of the diameter of stems, fruits, and roots. From the beginning of the 1990s, most authors working on irrigation scheduling have used linear variable differential transformers (LVDT), also called linear variable displacement transducers. The LVDT-type sensors are robust and of high precision, close to ±1mm. In many cases, however, the resolution cannot be expected to be greater than about 10mm, due to errors associated to calibration, voltage recording, temperature changes, etc.

Depending on the dendrometer model and the purpose of the measurement, nails or screws may have to be driven into the tree, either to support the instrument or to serve as a fixed reference against which growth is measured. This disturbance frequently evokes abnormal growth acceleration in the vicinity of the injury. For this reason, the dendrometer holder is usually attached tightly to the stem by elastic straps, which do not affect growth. Holders are usually of aluminium and INVAR, an alloy containing 64% Fe and 35% Ni, with a thermal expansion close to zero.

The dendrometers must be calibrated prior to installation. This can be done using a precision micro meter or metal plates of known thickness. In fruit trees and other plants of large stem diameter, dendrometers are installed on the side of the stem opposite to the sun's trajectory, to minimise negative effects of heating by direct solar radiation. They must be at a certain distance from the ground to avoid interference from growing weeds, and as far as possible from any scars, such as those caused by grafts, pruning, and tillage operations. The outer, dead tissues of the bark must be removed before installation, allowing the contact point of the sensor to rest directly on the living tissues of the bark. The dendrometer must always be in contact with the surface, for whicha spring or glue can be used, and the whole area must be covered with insulating materials (rubber foam, vaseline, silicon resin, etc.) and reflective thermo-protecting foil to minimise both heating by direct solar radiation and the impact of rain drops. Once installed, the dendrometer is connected to a data logger programmed to automatically scan the sensor outputs every few seconds and store average values every few minutes. The data logger is usually provided with a system for data transmission to the user's computer.

TDV Based Indices

The daily TDF cycle provides three different common indices; maximum daily trunk diameter (MXTD), minimum daily trunk diameter (MNTD) and maximum daily trunk shrinkage (MDS), the last calculated as the difference between MXTD and MNTD (Fig. 4.6). Trunk growth rate (TGR) can be calculated taking into account values of MXTD on two consecutive days (Fig. 4.6) or over longer periods of time in order to reduce the day-to-day trunk growth variability.

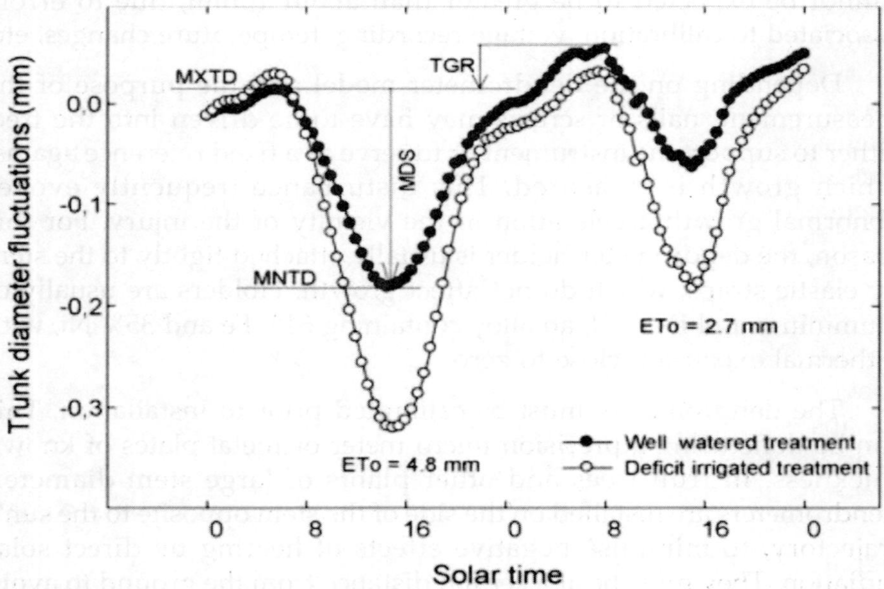

Fig 4.6 Trunk diameter variation based indices

Absolute indices

MNTD: Minimum daily Trunk Diameter. The minimum value of daily trunk diameter, which is reached sometime during the afternoon (mm)

MXTD: Maximum Daily trunk Diameter. The maximum value of daily trunk diameter, which is reached just before sunrise (mm)

Derived Indices:The maximum daily shrinkage (MDS) and stem growth rate (SGR) are, in that order, the most widely used SDV-derived indices.

MDS: Maximum Daily trunk Shrinkage. The difference between the maximum daily trunk diameter value (just before sunrise) and the minimum daily trunk diameter value (sometime during the afternoon) (mm)

TGR: Trunk Growth Rate. The difference between two consecutive maximum daily trunk diameters (mm).

Signal Intensity

The magnitude of stem contraction even in a well-watered tree is not constant over a period of days with different environmental conditions. Hence the absolute MDS values registered without considering the evaporative demand might be meaningless For this reason, for irrigation scheduling it is better to use the concept of signal intensity(SI), normalizing the absolute values with respect to values in non-limiting soil water conditions. The signal intensity for MDS was calculated as :

$$S_{1-MDS} = \frac{\text{Actual MDS}}{\text{Re ference MDS}}$$

MDS signal intensity is a dimensionless variable, where values above unity indicate water stress levels, while SI values of unity indicate the absence of irrigation-related stress (Goldhamer and Fereres, 2004). In addition, SI values clearly below unity could be used to predict the presence of trees suffering very severe water stress.

Signal : noise ratio

The wide tree-to-tree variability in TDF-derived parameters regardless of the irrigation programme is well documented. Consequently, for measurements to be accurate, it would be necessary to measure TDF in many trees within an orchard, increasing the measurement costs in the process. To assess the usefulness of plant-based water stress indicators for irrigation scheduling, it is important to characterize the signal:noise ratio, defining noise as the coefficient of variation of the mean (CV). Therefore, the uncertainty associated with a high noise value may not be as crucial when the signal strength is sufficient (Goldhamer and Fereres, 2001). The variability, or noise, of SI-MDS was quantified by the coefficient of variation (CV). The sensitivity was calculated as the signal:noise ratio:

Trunk diameter variation (TDV) is a better water stress indicator and it is possible to detect the plant water stress using TDV. TDV based indicator MDS SI can be successfully used for deficit irrigation scheduling of Lemon and Peach. The deficit irrigation protocol based on MDS SI was able to reduce seasonal water applied in peach trees by 35-42 %. The studies also showed that the irrigation protocol can be suitably modified to change water stress level as per growth stages by varying the MDS signal intensity. Since there was not a single study on most popular Indian fruit crops, the preliminary assessment of TDV under deficit irrigation should be done for Indian fruits e.g. Mango, Guava, Sapota, Litchi and Banana.

□□□

5

Precision Agriculture and Geographic Information System

Precision agriculture is an integrated agricultural management system incorporating several technologies including GIS and remote sensing for managing each crop production input like variety, seed, fertilizer, herbicide, insecticide etc. on a site with specific basis to reduce waste, increase profits and maintain the quality of the environment. Protected cultivation is a form of precision farming wherein intended with precise level of control over plant microclimate to manage stresses, specifically abiotic stresses for optimum plant growth and production. The microclimatic parameters being referred here are temperature, light, relative humidity, air composition and the nature of rooting medium. In open field conditions, it is not possible to control the environmental factors such as extreme of light during summers, freezing temperature during winters, rains, frost, hails and drought conditions. The climatic fluctuations and adversities cause poor crop productivity and/or sometimes, total crop failures.

Protected cultivation of horticultural crops offers the best choice for diversification from traditional agriculture production system for a number of reasons. Production of crops under protected conditions has great potential in augmenting production, productivity and quality of vegetables, flowers and some fruit crops during main and off-season and maximizing water and nutrient use efficiency under varied agro-climatic conditions of the country. This technology has a great potential especially in hills, plateau region and even in peri-urban areas, where land can be utilized judiciously for getting higher returns by growing high value vegetable crops (tomato, cherry tomato, coloured capsicum, parthenocarpic cucumbers), cut flowers (rose, carnation, gerbera, lilium) and fruits like strawberry, grapes etc. and also their healthy and virus-free seedlings. So, there is need of such technologies in the field of horticulture for feeding the rapidly increasing population and shortening the crop productivity gap of the country.

Advantages of Precision Farming cum Protected Cultivation

The precision farming cum protected cultivation technology plays significant role in managing natural resources in the hilly and plateau area in following manner:

- Control soil erosion and evaporation loss.
- Water saving through drip irrigation as compared to flood irrigation.
- Moisture conservation and nutrient leaching control.
- Less incidence of pest and diseases in protected agriculture.
- Enhance earliness and crop duration.
- Increased yield and quality.
- Higher productivity per unit area therefore, save land for others crops.
- Play a key role in INM, IPM and Organic food production.
- Promote Hydroponics / Aeroponics/ Aquaponics technology.
- Can be used on roof with pot cultivation.
- Beneficial for marginal or weaker section farmers
- Increased employment for livelihood.

Table 5.1 Current status of protected agriculture in India

S.No.	State	Area under protected agriculture (ha)
1	Maharashtra	8000
2	Karnataka	3000
3	Gujarat	2500
4	NEH State	800
5	Himachal Pradesh	500
6	J&K	400
7	Uttarakhand	300

Structures under protected cultivation

The protected horticulture includes various structures such as poly-house, poly-tunnel, shading net house, floating cover, bird proof net house, cloches, insect proof net house, mulching and anti-hail net house.

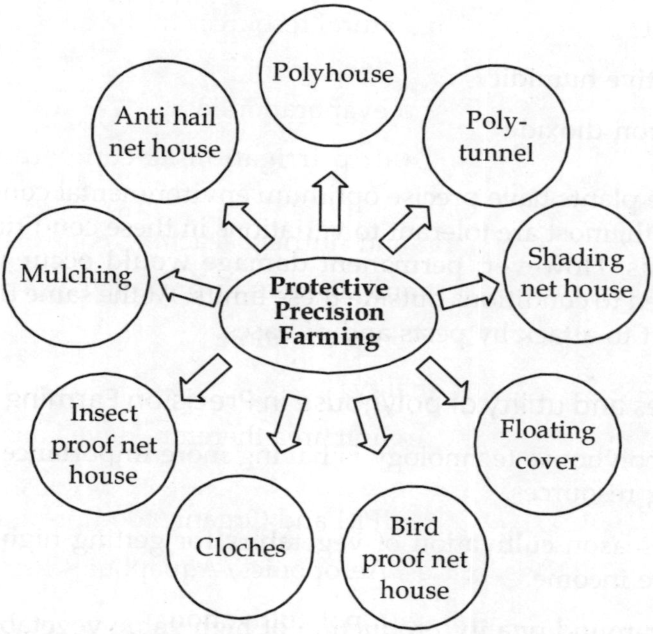

Fig. 5.1 Structures of protected Cultivation in Precision Farming Polyhouse

A polyhouse is quasi-permanent structure, covered with a transparent or translucent material, ranging from simple homemade designs to sophisticated pre-fabricated structures, wherein the environment could be modified suitable for the propagation or growing of plants. Materials used to construct a polyhouse frame may be wood, bamboo, and steel or even aluminum. Coverings material is generally plastic film.

Types of polyhouse

1. Naturally ventilated in different design
2. Semi-controlled or Fan-Pad or evaporative cooled in different design

Plant environment and polyhouse climate

A plant grows best when exposed to an environment that is optimal for that particular plant species. The aerial environment for the plant growth can be specified by the following four factors:

i) Heat or temperature

ii) Light

iii) Relative humidity

iv) Carbon dioxide

While plants have precise optimum environmental conditions for best growth, most are tolerant to variations in these conditions within some limits. However, permanent damage would occur when they are exposed to conditions outside these limits. At the same time, plants are subject to attack by pests and diseases.

Objectives and utility of polyhouse in Precision Farming

The polyhouse technology is having more importance regarding managing resources.

- Off-season cultivation of vegetables for getting high yield and more income.
- Year-round quality production of high value vegetable crops.
- Permanent plastic structures are suitable for high value off-season cultivation.
- Minimize crop stresses during growing season.

Integrated nutrient management (INM)

The integrated nutrient management (INM) for cultivation of cucurbits in poly-tunnel requires green manure and FYM/Compost @ 200q/ha mixed in the soil at the time of field preparation. The dose of NPK @ 40, 20, 20 kg/ha given three times a month for one crop duration through drip irrigation. The integrated nutrient management (INM) combination for cultivation of capsicum in poly-house requires green manure and FYM/Compost @ 400q/ha and it is mixed in the soil at the time of field preparation. The dose of NPK @ 140, 100, 120 kg/ha three times a month given through drip irrigation. Calcium is applied

@ 5kg/ha. The integrated nutrient management (INM) combination for cultivation of tomato in polyhouse requires green manure and FYM/ Compost @ 500q/ha and it is mixed in the soil at the time of field preparation. The dose of NPK @ 150, 120, 130 kg/ha three times a month through drip irrigation should be given. The application of calcium @ 5 kg/ha and boron @ 3-4 sprays should be done. The integrated nutrient management (INM) combination for cultivation of cucumber in polyhouse is FYM/Compost @ 300q/ha and it is mixed in the soil at the time of field preparation. The dose of NPK @ 50, 30, 40 kg/ha through drip irrigation should be given. The application of calcium @ 3 kg/ha should be done. The integrated nutrient management (INM) for cultivation of summer squash in polyhouse require FYM/Compost @ 200q/ha and it is mixed in the soil at the time of field preparation. The dose of NPK @ 40, 20, 30 kg/ha through drip irrigation should be given.

Table 5.2 Suitable Varieties of crops for precision farming

Crop	Variety
Capsicum	Bharat, Mahabharat, Indam-5, Tanvi, Orobelle, Swarna, Bomby, Indra, Nun-3029
Tomato	Rakshita, Naveen 2000, Nutan, Snehlata, Avinash-2 and GS-600
Cherry tomato	Pusa Cherry-1, NS2
Cucumber	Kian, Isatis, and Hilton
Summer squash	Himanshu, Seole green, Dokato, PusaAlankar and Australian green

Poly-tunnel in Precision Farming

Poly-tunnels are flexible transparent coverings that are installed over single or multiple rows of vegetables to enhance the plant growth by warming the air around the plants in the open field during winter season when the temperature is below 8^0C. Plastic low tunnels are often used to promote the growth of plants during the period of winter season. Low tunnels are supported above the plants by using loops of GI wire and a clear or transparent plastic of 20-30 micron is covered/ stretched over the loops and the sides are secured by placing in soil. Plastic is vented or slitted during the growing season as the temperature increase within the tunnels.

Fig. 5.2 Cultivation in Poly tunnels

The farmers can grow different varieties of summer squash (round-fruited, long-fruited), which is a emerging crop along with cultivation of netted muskmelon varieties in place of traditional varieties. Bitter gourd and round melon are two other crops with increasing demand and which usually fetches very high price during off-season and can be grown successfully by using the plastic low tunnel technology. This technology is highly suitable and profitable for the farmers.Poly-tunnel is also known as temporary or mobile micro-polyhouse or row-cover and classified as low poly-tunnel, high poly-tunnel, and trench poly-tunnel.

Objectives and utility of Poly-tunnels

The poly-tunnels are of same importance as poly-house and are used for:

- Off-season cultivation of vegetables for earning more income
- Walk-in tunnels are only erected over the crop during the peak winter months of December-mid February and thereafter, structure is removed from the crop.
- Since, the plastic is used only for two months (December-mid February) therefore; life of the plastic can be 8-10 years.
- These temporary plastic structures are suitable for off-season vegetable cultivation in northern plains and low hills.

Shade-Net Cultivation in Precision Farming

Shading nets are the perforated plastic materials used to provide relief to the plants from the scorching sunlight, hot winds, direct rainfall as well as insects-pests. These shading nets are available in different shading intensities ranging from 25% to 75%. Theses shading nets are being used for raising nursery, indoor plants, hardening of tissue cultured plants and growing of vegetables. During high light intensities, leafy vegetables and ornamental greens are recommended to be grown under shade nets. This structure enhance 25-35% more yield as compared to traditional methods during peak summer season.

Shade-Net

Insect Proof Net-house in Precision Farming

Shade house and Insect proof net house are often synonymously used but more correctly a net house is enclosed with perforated screen primarily to act as a barrier for the entry of insects and pests. Insect-proof nylon nets are available in different intensities of perforations, ranging from 25 mesh to 60 mesh. Nets of 40 or higher mesh are effective means to control entry of most flying insects and save crop from diseases. These structures permit early planting of tomatoes and capsicum without the risk of vector. Higher mesh size, however, reduces the air exchange of the structure. UV- stabilized nets are now available which have a longer life. This structure enhances 35-40% more yield as compared to traditional methods.

Net-house

Mulching in Precision Farming

The Mulching technology is most important component of protected agriculture and classified in two groups:

- Bio-mulching
- Plastics mulching

Mulching is the practice of covering soil around plants with an organic or synthetic material to make conditions more favorable for plant growth, development and crop production. The change in soil environment could be favorably controlled by selection of an appropriate mulching material. Bio-mulches such as leaves, straw, saw dust, peat moss, compost and gravel have been used for centuries to control weeds, hold moisture and conserve soil. More recently, paper, petroleum and plastics mulches have been tested for soil environment control. The most versatile of all the mulches at present is, however, plastics mulch which is manufactured in different film colors, thickness and widths. These plastics film mulches are generally made from low density polyethylene or linear low density polyethylene. By a proper selection of plastics mulch composition, color and thickness, it is possible to precisely control the soil environment. Black is the most common color used for mulching. Transparent mulch film increases the soil temperature. The optimum thickness of polyethylene should be around 40-50 micron.

Advantages of mulching

- Conserves soil moisture and reduces nutrient leaching
- Improves micro-climate around plant root zone
- Minimizes weed incidences
- Minimizessoil erosion and evaporation losses
- Saves 40-50% irrigation water, 20-25% fertilizers
- Increases 30-40% yield
- Continuous uptake of nutrient
- Decomposes organic matter and improves soil quality
- Minimizes 80-90% weed problem and weeding charge
- Controls 80-90% forest pruning for using tomato and cucurbits staking.

Quality Nursery Raising Technique

Plastic has wide variety of uses in nursery *viz.* material glazing, seeder, trays and cell pack for propagation, container carrying trays, container for propagation and labels for proper identification, ground cover and packing of planting material for long distance transportation. Generally plastic trays or pro-trays having different cell sizes are used for raising vegetable seedlings. Mainly two kinds of plastic pro-trays are used in raising vegetable seedlings. One type of tray has cavities of 3.75 cm size, whereas the other type has cavities of 2.5 cm size which are used based on seed size and root growth. These trays helps in proper germination of seeds, provide independent area for each seed to germinate, reduces the mortality rate, maintain uniform uniform and healthy growth of seedlings, are easy in handling and storing, reliable and economical in transportation. These plastic trays may be fixed in thermocol base trays having the same number and size of cavities before filling the media.

Drip Irrigation

Drip irrigation is the best available technology for precision farming which is the judicious use of water for growing horticultural crop in large scale on sustainable basis. Drip irrigation is a low labour intensive and highly efficient system of irrigation, which is also amenable to use in difficult situations & problematic soils, even with poor quality water. Irrigation water savings ranging from 36-79% can be affected by adopting a suitable Drip irrigation system. In drip irrigation, water is supplied through a network of plastic pipes using dripper/emitters. Water is supplied at a regular interval and at a required time/ quantity. Productivity gains vary from 20 % to 50 % depending upon the crop. Drip irrigation works by applying water slowly, directly to the soil. Drip irrigation has high efficiency as there is no loss of water in form of evaporation or run- off and water is applied directly to the plant roots.

Drip Irrigation

Materials used for making protected structures

Various material are used in protected structures for installation and covering such as G.I. pipes, transparent film, shade net, mulching sheet, drip irrigation system, insect-proof net, pro-trays, plastic ropes, angel iron etc.

- 200 micron/190 GSM U.V. stabilized transparent film
- 30-50% U.V. stabilized different colours agro-shade-net
- 40-50 mesh/sq. inch U.V. stabilized white colour insect-proof nets
- 2040 micron/10-30 GSM U.V. stabilized different colour mulching sheet
- MIS for irrigation and fertigation
- Pro-tray, pot, plastics ropes etc.
- Bamboo, Wooden pillar, Stone pillar, Angel Iron, G. I. Pipe, G. I. Wire, Nut-bolt, building materials etc.

Constraints for Success of Precision Farming in Horticultural Crops

There are certain constraints for success of managing resources in protected agriculture such as infestations of soil & air borne diseases, poor quality of soil health, water and traditional methods of hill and plateau agricultural practices. It requires high initial input cost to make use for protected cultivation.

The precision farming cum protected cultivation technology, due to higher yield per unit area along with export quality produce, can prove useful in raising the socio-economic status of small and marginal farmers of the hilly and plateau areas. It is also important in present scenario of shrinking cultivable land, depleting water and other natural resources. Keeping in view the increasing demand of off-season and high value vegetables, fruits and flowers in metropolitan cities of the country, there is urgent need for diversification of traditional agriculture by production of high value horticultural crops under different protected structures for increasing the productivity and quality produce for getting high returns. Nursery raising under protected cultivation can be adopted well as an agri-entrepreneur in major vegetable growing areas of the country by unemployed youths who are graduate in agriculture or post graduate in horticulture. Low-cost protected technology like plastic low tunnels or walk in tunnels can be used for off-season vegetable cultivation for getting high returns. Similarly insect-proof net houses can be used on a large scale for safe vegetable cultivation to minimize use of pesticides in vegetable cultivation and virus free crop production. Protected cultivation of horticultural crops is reckoned to be the next logical step to open field agriculture. One million hectare under protected vegetable cultivation can easily increase the production of vegetables by about 40 million ton. This will also provide ample opportunity for the creation of additional jobs in rural sector.

In India, the field of precision farming cum protected cultivation is still evolving. Methods of more efficient input application and protection technology are being devised. More eco-friendly plastics are being produced for making the processes more efficient. As a result, protected cultivation could bring prosperity to horticultural crop growers in hilly and plateau regions of India and ensure nutritional security for everyone.

Geographical Information System

Precision agriculture (PA) is a farming management concept based on observing, measuring and responding to inter and intra-field variability in crops. There is a lot of technology used to make modern agriculture more efficient. Use global positioning systems (GPS) and GPS-computer guided tractors and harvesters are the major technologies in mapping the soil characteristics, crop growth and yield is increasing at faster rate. The technology PA can be used in electromagnetic soil mapping, crop yield data collection, aerial imagery, soil types, soil characteristics, which will help in taking decision on of spatially targeted inputs and other field level management practices. Each of these geo-referenced data layers helps subdivide a large field area into smaller management zones. Using small management zones reduces waste while increasing production potential.

One example of a precision agriculture practice is to evaluate the natural soil variability of a field. If the soil in one area holds water better, crops can be planted more densely and irrigation can be sparing. Or, if the plot is used for grazing, more cattle can graze than a similar area of poorer quality soil.

By studying these factors and using precision agriculture, farmers are able to produce more food at a fraction of the cost. Farmers also conserve soil for sustainable food production. Precision agriculture results in a stable food supply, which results in a strong community.

Remote Sensing (RS) is the technique of obtaining information about an object on the earth through the analysis of data acquired by a device that is not in contact with the object. Remote sensed data or image can be obtained by using various sensors. This works on the principle of spectral reflectance signature.

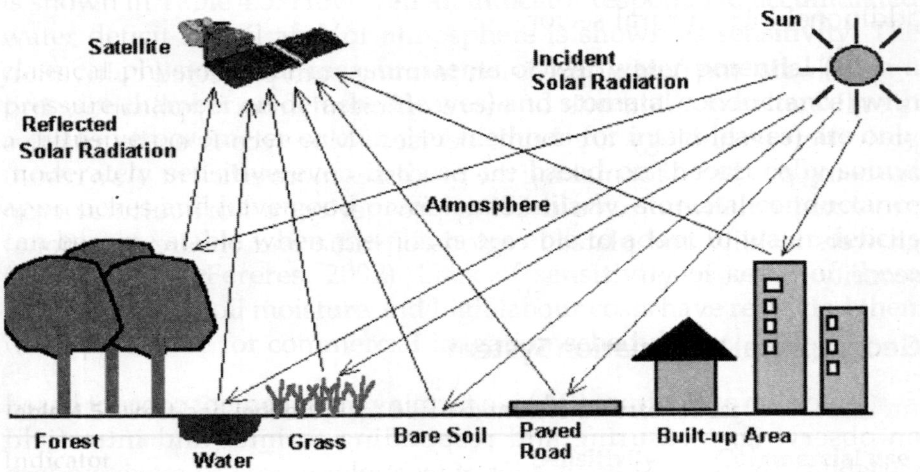

Fig. 5.3 Sensing spectral reflectance using satellites

Common types of sensors used for acquisition are LISS III, LANDSAT TM/ ETM, MODIS, AVHRR, SAR, RISAT, AWIFS, SPOT, IKONOS, etc. Remote sensing data are generally collected in the raster format (Lee, 1991). Raster images are also called bitmap images. Raster images have a finite set of digital values called picture elements or pixels. Pixels are the smallest individual element in an image.

Geographic Information System is a system of hardware, software and procedure designed to support the capture, management, manipulation, analysis, modeling and display of spatial data. GIS and remote sensing are well established information technologies for land and natural resources management and are now widely recognized (Zhou, 1995). GIS can show many different kinds of data on one map. This enables easy analysis of data to understand patterns and relationships. With the help of GIS, locations of different things on the earth such as water bodies, natural vegetation, agricultural land, settlement, etc. may be identified. So, through remote sensing the acquisition of spatial data, GIS for undertaking integrated analysis, presentation of spatial and associated attribute data are found to be much more effective for getting the information about land use and land covers (Lillesand and Keifer, 2001).

Land use and land cover

Land use and land cover are two different terminologies that they are very often used as a substitute to each other (Dimyatri *et al*, 1996). Land cover refers to the physical materials on the surface on the earth viz. natural vegetation, barren land, water bodies, etc. while land use refers to the human activities that takes place on or make use of land such as, residential, industrial, commercial, agricultural land, etc. (Longley *et al.*, 2001).

Land use planning for agricultural development using remote sensing and GIS

Remote sensing and GIS are widely used technologies for the land use planning for agricultural development.Growing population pressure and human activities are increasing the demand on the limited land resources, both for agricultural and other uses. To meet this unprecedented demand of land for various activities, reliable and comprehensive information about the spatial distribution of land use and land cover categories and their present status is a prerequisite for planning, utilization and management of land resources of any region such as for example eastern plateau and hill region. This type of spatial information about land surface may be obtained with the help of remote sensing and GIS techniques. Land is the most important natural resources which contains soil, water, vegetation on which all human activities are based. Land use planning has been defined as the systematic assessment and management of land and water resources for their optimum utilization for increasing the agricultural production. In eastern plateau and hill region, the land use practices like agriculture, plantation, agro-forestry, agro-horticulture, afforestation, dry land horticulture may be done for the development of agriculture in this region.

Land use and land cover classification

Multispectral image classification is the process of sorting of pixels into a finite number of individual classes or categories of data based on their digital number (DN) values. Digital number is also known as pixel value or spectral reflectance value. Image classification is an important part of remote sensing, image analysis and pattern recognition. Land use and land cover classification of satellite images is an important activity of extracting geospatial information for the development of land use planning such as agricultural development, township planning, etc. The most common methods of image classification are supervised and unsupervised. Supervised classification is closely controlled by the analyst/ user. In this process, pixels that represent patterns or land cover features that are recognized or identified with the help of other sources such as aerial photo, ground truth data or maps. Knowledge of data and classes desired is required before classification. By identifying patterns, computer system is instructed to identify pixels with similar characteristics. Minimum distance to mean, parallelepiped and Gaussian maximum likelihood algorithms are used in the supervised method of image classification.

Unsupervised classification is more computer oriented. It enables to specify some parameters that the computer used to uncover statistical patterns that are inherent in the data. These patterns don't correspond to directly meaningful characteristics of the scene, such as contiguous, easily recognized areas of particular soil type or land use, they are simply clusters of pixels with similar characteristics. Unsupervised training is dependent up on the data itself for the definition of classes. This method is usually when less is known about the data before classification. The analysis responsibility, after classification, to attach meaning to the resulting classes whereas in supervised training, classes are predefined. Thus the classes are pre-defined in supervised classification but in unsupervised classification, classes are defined after classification. ISODATA (Iterative Self Organizing Data Analysis Techniques) clustering and C-means clustering techniques are generally used in unsupervised classification method. With the help of image classification, agricultural planning may be done for the development of agriculture. LISS III image of Ranchi area has been collected and georeferenced according to Survey of India (SOI) toposheet (Fig 5.4).

LISS III image of the Ranchi was georeferenced and was classified using maximum likelihood technique taking natural vegetation, agricultural land, standing water bodies, built-up area as different classes or categories. The classified image showing delineated fallow land is presented below (Fig 5.5).

Comparison of classified images at two different times can be used to assess the changes in areas of agricultural land, fallow land, water bodies, natural vegetation, cropped area and built-up area. This spatial data may be useful in the formulation of agricultural planning in this area to increase the agricultural productivity.

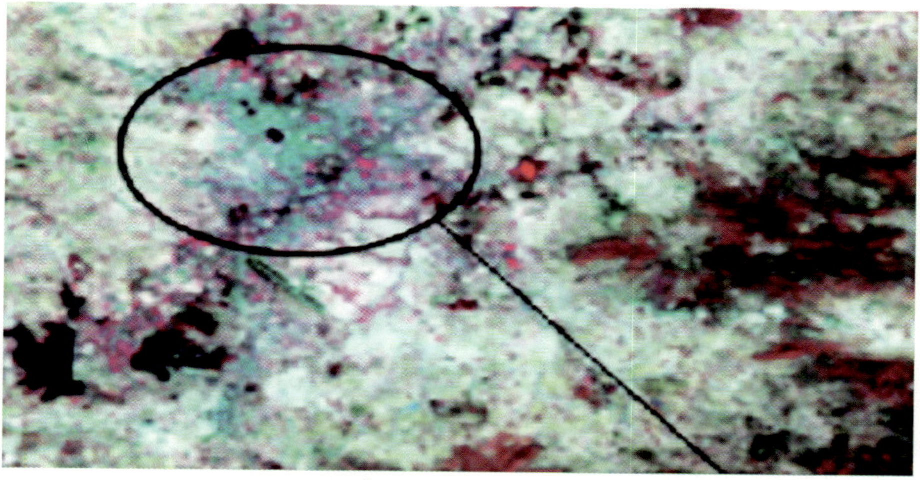

Fig. 5.4 LISS III image of Ranchi area

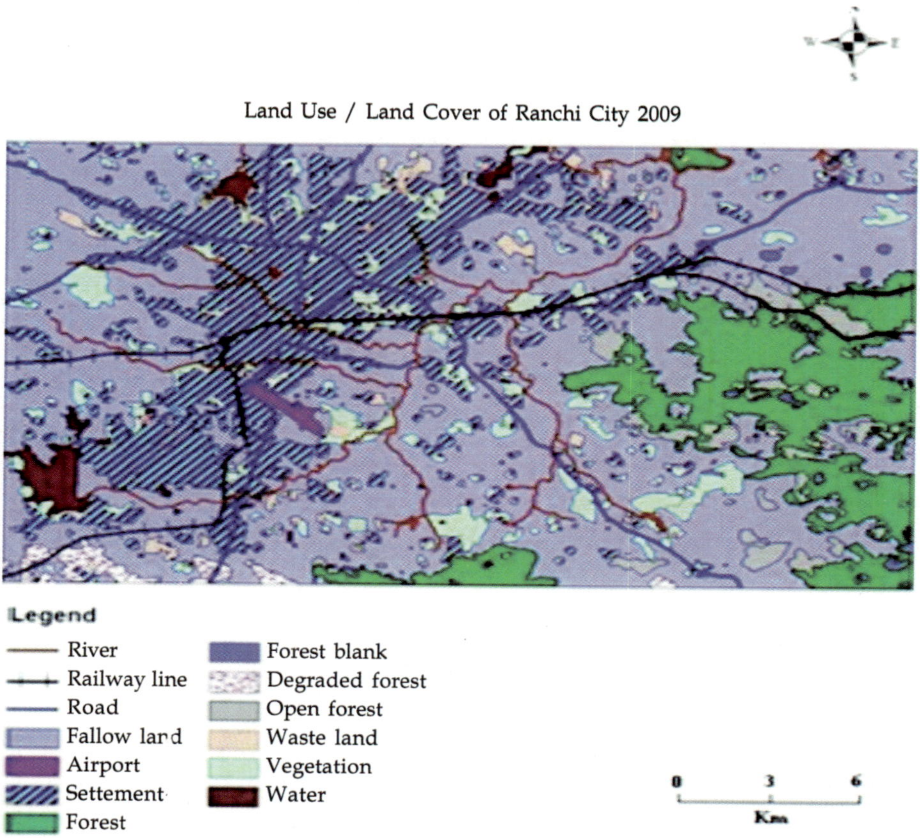

Fig. 5.5 Classified LISS III image of Ranchi area

Remote sensing and GIS techniques have been found to be the most important tool for the assessment, mapping, management and planning of natural resources such as soil, water, vegetation, etc. for the development of agriculture. With the help of RS and GIS, different land covers such as location of soil, water, vegetation, etc. may be identified and mapped and this information may be used for planning and development of agriculture in any area including hill and plateau region and for the development of built-up areas also.

❑❑❑

6

Fruit Based Production Systems

With increasing population, the pressure on agricultural land for non-agricultural use is increasing day by day. Under this scenario, increasing productivity of different agricultural production system from the present level of productivity is the need of the hour. Again, gradual decrease in profitability of different traditional agricultural practices due to decrease in the land holding size and non-adoption of modern agricultural technological options is resulting in gradual shift in the dependency on agriculture to non-agricultural occupation and increased migration of rural people to urban area with poor living conditions. Under this scenario, there is urgent need for adoption of technological options for increasing the productivity and profitability of different agricultural enterprises.

In most part of the East Indian plateau, the low lands and the medium uplands are used for the rainfed rice production whereas the uplands are seldom utilized for production of agronomical crops. These uplands are either used as pasture or for social forestry. To ensure high returns from the underutilized and stressed lands and improve the soil characteristics, perennial fruit plant based production system have been found successful under semi-arid and dry land conditions (Raturi and Chadha, 1993). Storing carbon (C) in vegetation and soils is one of the strategies accepted by the United Nations for mitigating high atmospheric concentrations of carbon dioxide (CO_2) that causes global warming. These multispecies plant associations are speculated to have high carbon sequestration potential (CSP) due to their forest-like structure and composition (Nair *et al.*, 2009), particularly in accumulating C in the soil. The ecological process by which trees contribute to such higher amounts of C in stable form is believed to be mediated via litter fall and decomposition, and root activity especially decomposition of sloughed-off roots (Lemma *et al.*, 2007).

Perennial crop based cropping system offer efficient alternative for improving the total factor productivity of these stressed lands. These cropping systems are dynamic interactive practices aimed at better use of the production components such as soil, water, air space, solar radiation and all other inputs on a sustainable basis. Subsistence farmers who face famine would consider a technology to be one that produces atleast some yield in the worst year. Subsistence productivity is difficult to achieve by growing only seasonal or forage crops. Under such situation adoption of fruit based cropping system seems an option to achieve these objectives. Fruit based cropping system is a planting system comprising combinations of plants with various morpho-phenological features to maximize the natural resource use efficiency and enhanced total factor productivity. High density orchards and multitier cropping systems are the technologies that offer enormous potential in improving land and water productivity in the region.

High Density Orcharding

Horticultural crops have played a major role towards diversification agriculture for ensuring nutritional and livelihood security of farmers. Over the last five years, the scenario of fruit crops in the country has changed drastically in terms of increased production. However, a major share of this increase in production is attributed to largescale increase in the area under different fruit crops. Contribution of increased productivity to this increased production is meager. In case of Bihar, there has been a increase in productivity fruit crops from 11.13 t/ha to 11.80 t/ha during last five years as compared to 10.0 to 15.1 t/ha in case of eastern Uttar Pradesh. Hence, there lies a large potential for increase in the productivity of fruit crops in different states.

Productivity has generally been evaluated on an individual tree basis in the past. However, now it is considered in terms of land area, reflecting a more efficient means of utilizing this basic resource that is becoming scarcer and costlier gradually. The productivity of a given land area is related directly to fruit bearing foliage and its photosynthetic capacity. These concepts have converged themselves in the form of high density orcharding in fruit crops. High density orcharding is an efficient option for increasing the productivity of different fruit crops. High density orcharding is the practice of accommodating higher number of plants of the same species per unit area by planting at a closer spacing than that is planted under traditional system. The principle of increased productivity through high density orcharding are increased sunlight capture per unit area, land use efficiency and appropriate vegetative reproductive balance of the plants. However it would be more purposeful to the orchardist as well as researchers if its definition is widened to include the accommodation of more than one species so that the maximum potential of a unit land area could be economically utilized for increased production.

In fact it is difficult to define high density planting in terms of number of trees alone. In a high density planting system the basic principle should be to plant at a density that will fill the space quickly and without mutual shading ultimately (Iyer and Kurian, 2006). In case of multiple species high density planting, the interception of solar radiation at different tiers by canopies of various species, based on their light transmission characteristics and shade tolerance are primarily exploited to enhance productivity and profitability. The high density planting may therefore be defined as a cropping system in which highest number of plants are accommodated within unit area so as to obtain maximum output by optimum utilization of solar radiation, land, water and soil nutrients.

In India, single species high density orcharding has been a success story in fruit crops like apple, banana, pineapple, papaya and mango where yield increases ranging from 1.5 times to 2.5 times have been recorded. Apart from yield advantages, high density orchards have several other advantages over the conventional low density planting. It facilitates better utilization of incident solar radiation and increase in bearing surface per unit land area. It is more amenable to modern input saving techniques like drip irrigation, fertigation, mechanical harvest as well as horticultural operations such as pruning and plant protection measures that enhances the labour output efficiency in addition to the increased effectiveness of these operations (Iyer and Kurian, 2006). Better fruit quality due to optimum exposure to sunlight and easier harvesting of fruits with less injury, resulting in better post-harvest life are added advantages of high density orchard.

Historical background

The first high density orchards of apple, particularly by hedgerow were established during 1962 in England and by 1977, its proportion in total apple acreage of the country increased to about 3%. In Italy, Switzerland, Austria and Yugoslovia, the first high density orchards with 1200 to 2800 apple trees per ha were established by the end of 1960's and a major increase in its area came during 1070's. Following the lead in these countries, high density orchards with 1250 to 5000 apple trees per ha and 1500 to 2500 pear trees per ha were established in countries like The Netherlands, Belgium and West Germany. Gradually interest started developing in many tropical regions as India on herbaceous perennial crops like banana and pineapple which later extended to woody perennials like mango, cashew and guava.

Strategies for high density orcharding in fruit crops

Principles of high density orcharding which have been developed over the years for temperate fruits can be attempted for woody perennials with certain modifications. There modifications are needed since it has to be remembered while attempting to restrict the size of tropical fruit trees that the evolutionary development of these trees in tropical forests where they originated, normally give rise to large trees whose survival depend upon vegetative vigour rather than prolific production of fruits and seeds An orchard has three spatial dimensions-length, width and height. Length and width are ground surface dimensions while height is determined by the height a grower allows his trees to reach within the confines of the natural cultivar-rootstock combination. Control of tree height is needed for allowing optimum interception of solar radiation by the foliage as well as easy and effective orchard operation.

Planting system and spacing: Ground dimensions are optimized for achieving higher planting densities through appropriate planting systems or orchard designs involving plant to plant and row to row spacing. In general different types of planting systems like square system, rectangular system, triangular system, hexagonal system, paired planting system and cluster planting system are followed in fruit orchards. Modern high density orchards are being raised by more intensive planting systems such as hedgerow system (single or double or multiple). Research work done at different parts of the world on effect of planting densities in different fruit crops have been summarized below.

Canopy management : The third spatial dimension of the orchard i.e. height can be managed primarily through training aided by pruning, vigour regulating rootstock, chemical growth regulators etc.

Training: Training is a horticultural practice in which tree growth is directed into a desirable shape and form. Training young trees is essential for proper tree development. It is better to direct tree growth with training during initial years of orchard establishment than to correct it with pruning at later stages. The goal of tree training is to direct tree growth to develop plant form that is good in light utilization, small in size and easy to manage especially in terms of mechanization and to minimize cutting and branch removal at later stages. An ideal tree form is one that captures maximum sunlight with minimum support portion of the tree. Conventional training systems such as the central leader and open centre training system can be employed for medium high density or semi intensive bush orchards while more intensive orchards require non-traditional systems like slender spindle plant form, thin restricted plane canopy or the columnar form.

Pruning:Pruning is the removal of a portion of a tree to shape or maintain tree structure and productivity. Pruning done for giving the trees a proper shape or form during the initial orchard years is termed shape pruning or formative pruning while that done to maintain tree structure, size and productivity during the later year is called maintenance pruning. Type of pruning and season of pruning have bearing on the tree vigour. A thinning cut removes an entire shoot or a branch back to its point of origin. Thinning cuts do not invigorate the tree. A heading cut or a tipping cut removes only the the terminal portion of a shoot or branch. This type of cut promotes the growth of several buds below the cut. When lateral branches are headed into one year old wood, the area near the cut is invigorated. The headed branch is much stronger and rigid, resulting in lateral secondary branching. Research work carried out on pruning in different fruit crops under high density planting is summarized below.

Use of plant growth regulators: Growth retardants can be successfully used to control vigor of fruit trees under high density orcharding. There are three basic mechanisms by which tree height can be restricted by chemicals. The most important chemicals of this group include Maleic Hydrazide, fluorenols and in some cases ethylene and ethylene releasing compounds. These chemicals, however are not much use in controlling tree size in commercial fruit production as they also inhibit normal leaf and flower initiation and development at the concentrations required to bring about growth inhibition in shoots.Reduction of tree height may also result from simultaneous growth of many shoot axes, which owing to competition for limited nutrients reduce the elongation of one or more axes on the same root system. Promising chemicals for inducing axillary bud growth include TIBA which is an inhibitor of auxin synthesis and translocation, cytokinins particularly benzyl adenine which promote growth of axillary buds. Unlike those of previous group, these chemicals are not translocated to the meristems of the new lateral branches and hence do not interfere with leaf and flower development.Near specific inhibition of internode elongation without concomitant disruption of apical meristematic function and loss of apical control is caused by certain growth retardants like SADH, CCC and paclobutrazol. These growth retardants reduce gibberellins biosynthesis, though different retardants act at different steps in the biosynthesis pathway. SADH prevent the conversion of trans-geranylgeranyl pyrophosphate to kaurene whereas paclobutrazol retards the formation of kaurene-19-ol from kaurene.

Use of dwarfing rootstock: Dwarfing rootstock are economical, effective and environmentally safe means for controlling tree vigour for high density orchards. Rootstocks for high density planting must control tree size, reduce vigour and induce precocity. Impact of dwarfing rootstocks upon the changing character of the fruit industry has been impressive in case of temperate fruit trees like apple. Several vigour restricting rootstocks such as M9 and M 26 are available for apple which have helped to plant the trees at very high densities. However, meager information is available on dwarfing rootstock for tropical and subtropical fruits. Vellaikulumban, Olour in mango, troyercitrange, Flying Dragon, Fortunella, Cuban Shaddock, Bouquet Sour orange, Thomsvillecitrangequat, *Feronia,* Rubidioux in citrus, Aneuploid 82, Chinese guava in guava are some of the promising dwarfing rootstocks in different fruit crops.

High density orcharding in mango

In mango, planting of plants of cv. Amrapali at a spacing of 2.5 x 2.5 m accommodating 1600 plants per ha can result in a yield level of 11.5 t/ha which can increase upto 24.46 t/ha in the 11th year. Due to the regular bearing habit of mango cv. Amrapali, the plants with regular dehorning (pruning), can be maintained at such closer spacing for atleast 10-12 years without intermingling of canopy. In case of popular but vigorous and alternate bearing mango varieties like Mallika, Langra, Bombay Green, Himsagar, planting at a spacing of five meters accommodating 400 plants per ha can also result in higher yield than the traditional system of orcharding by dehorning (pruning) of bearing branches to 60% of shoot length every alternate year.

Laying out and management of high density mango orchard:

- For laying out of high density orchard of mango, laying out of field should be done in the month of March to April and pit digging should be completed by mid of May.
- During laying out, care should be taken for proper alignment of the lines by use of techniques like 'drawing of right angled triangle of size 3 m x 4 m x 5m.
- For digging of pits of size 3 ft x 3 ft x 3 ft, the soil of upper half of the pit should be kept separately than the lower half of the soil. In case of high density orcharding of mango cv. Amrapali, planted at a spacing of 2.5 m x 2.5 m, the pit size can be reduced to 2 ft x 2 ft x 2 ft. The pits should be kept open for atleast one month during the summer.
- During last week of June, the pit should be filled up. While filling up the pits, the top soil should be mixed with 10 kg of *Trichoderma* + *Pseudomonas* treated FYM/Vermicompost, 2 kg of ground Karanj/Neem cake and 500g Single Super Phosphate or 300 g DAP. The mixture should be filled up in the pit first and the leftover space in the pit should be filled up with the bottom soil. Filling should be done in such a way that the soil layer remains above the top of the pit.
- For preparing *Trichoderma* + *Pseudomonas* treated FYM/ Vermicompost, 1 kg of *Trichoderma* + *Pseudomonas* formulation should be thoroughly mixed with 500 kg of FYM/Vermicompost. The mixture should be left on ground as heap for atleast 15 days. For maintaining proper moisture level of the mixture, the heap should be sprinkled with water and covered with polythene sheet. The mixture should be stirred twice before use.
- Planting can be done during July after 2-3 rains. It helps in settling of the soil in the pit. In case of area with water stagnation during monsoon, the planting can be done in the month of November with assured irrigation and proper winter protection.

- For planting, mango saplings of, care should be taken for obtaining planting material from genuine sources. The selecting proper quality sapling, it should be noted that the saplings have been prepared in polythene bags and height of sapling is not more than 90 cm. It should also be noted that the graft union is atleast 10 months old.
- During planting care should be taken to keep the graft joint above the ground. After planting, the soil around the plant should be thoroughly compressed.
- After planting, the plant should be irrigated. For preventing damage due to termite attack, termicide like Chloropyriphus @ 2 ml/ltr should be applied.
- Staking of plant is essential to prevent lodging of the plant. For this H-shaped stake should be prepared. For preventing damage of stake due to termite attack, the lower portion of the stakes should be dipped in strong termicides like Bifenthrin before putting in soil.
- Mulching the plants with grasses or straw after rainy season helps in conserving moisture in the root zone. For preventing attack of termite on mulching material, termicides should be used atleast 6 times during first year of plantation.
- Use of drip irrigation not only helps in effective use of water in the orchard but also minimizes labour requirement for irrigation. In case of orchards where drip irrigation cannot be installed due to any reason, care must be taken to irrigate the plants at regular interval. During winter season, the one year old plants should be irrigated twice a week and during summer season the one year old plants should be irrigated thrice a week.
- For training of the plants, care should be taken, not to allow any branches till height of 45-60 cm. After topping of plants, care should be taken to retain three well spaced shoots to act as primary branch. The other shoots should be removed. two to three months (October-November) of planting, the plants should be topped at a height of 30-40 cm. After 15 days of topping new shoots will emerge. The emerged shoots should be pruned to 50% of their total length after 3-4 months (February-March) so that new shoots will emerge below the cut point. The emerged shoots should again be pruned to 50% of their total length after another 3-4 months

(May-June). This will in development of proper canopy and initiation of fruiting from the first year of planting.

Ultra high density orcharding in guava

Under traditional system of guava orcharding guava is planted at a spacing of 5 m x 5 m accommodating 160 plants per acre. With optimum fruiting, a maximum yield level of 9.6 tonnes per acre can be obtained from a traditional guava orchard of more than 10 years of age. This low level of fruiting under traditional guava orcharding, makes it an unattractive venture for farmers of the plateau region. Being a current season bearer, guava plants respond to pruning. This behaviour of guava plants makes it an ideal plant for ultra-high density orcharding. Ultra-high density orcharding is a method of planting the saplings at a closer spacing in such a manner that more number of plants can be accommodated per unit area for obtaining higher yield per unit area than that over traditional system of orcharding. Being responsive to pruning, heavy bearing guava plants with short stature and compact canopy can be obtained through regular pruning. Keeping this in view, the 'Ultra-high Density Guava Orcharding' have been developed at ICAR Research Complex for Eastern Region, Research Centre, Ranchi with financial assistance from NABARD. In this system of orcharding, guava plants are planted at a spacing of 1 m x 2 m accommodating 2000 plants per acre. Under this system of orcharding a yield level of 12 tonnes per acre can be obtained during the third year and a yield level of 20 tonnes per acre can be obtained from 5 year old orchard.

Establishment and Management

- For establishment of ultra-high density guava orchard, the laying out should be initiated during March-April.
- For laying out of the orchard, marking should be done at a spacing of 1 m x 2 m distance. For obtaining perpendicular rows, the method involving right angled triangle of arm size 3 units x 4 units x 5 units should be used.
- During the month of May, pits of size 2 ft x 2 ft x 2 ft should be dug. While digging the pits, the upper half soil should be kept separately from lower half soil layer. The pits should be kept open for atleast one month during the summer.

- During last week of June, the pit should be filled up. While filling up the pits, the top soil should be mixed with 10 kg of *Trichoderma* + *Pseudomonas* treated FYM/Vermicompost, 2 kg of ground Karanj cake and 300g Single Super Phosphate. The mixture should be filled up in the pit first and the leftover space in the pit should be filled up with the bottom soil. Filling should be done in such a way that the soil layer remains above the top of the pit.
- For preparing *Trichoderma* + *Pseudomonas* treated FYM/ Vermicompost, 1 kg of *Trichoderma* + *Pseudomonas* formulation should be thoroughly mixed with 500 kg of FYM/Vermicompost. The mixture should be left on ground as heap for atleast 15 days. For maintaining proper moisture level of the mixture, the heap should be sprinkled with water and covered with polythene sheet. The mixture should be stirred twice before use.
- Planting can be done during July after 2-3 rains. It helps in settling of the soil in the pit.
- For planting, guava saplings of cv. Allahabad Safeda or Sardar should be used. Care should be taken for obtaining planting material from genuine sources.
- During planting care should be taken to keep the graft joint above the ground. After planting, the soil around the plant should be thoroughly compressed.
- After planting, the plant should be irrigated. For preventing damage due to termite attack, termicide like Chloropyriphus @ 2 ml/ltr should be applied.
- Staking of plant is essential to prevent lodging of the plant. For this H-shaped stake should be prepared. For preventing damage of stake due to termite attack, the lower portion of the stakes should be dipped in strong termicides like Bifenthrin before putting in soil.
- Mulching the plants with grasses or straw after rainy season helps in conserving moisture in the root zone. For preventing attack of termite on mulching material, termicides should be used atleast 6 times during first year of plantation.
- Use of drip irrigation not only helps in effective use of water in the orchard but also minimizes labour requirement for irrigation. In case of orchards where drip irrigation cannot be installed due

to any reason, care must be taken to irrigate the plants at regular interval. During winter season, the one year old plants should be irrigated twice a week and during summer season the one year old plants should be irrigated thrice a week.

- For training of the plants, care should be taken, not to allow any branches till height of 30-40 cm. After two to three months (October-November) of planting, the plants should be topped at a height of 30-40 cm. After 15 days of topping new shoots will emerge. The emerged shoots should be pruned to 50% of their total length after 3-4 months (February-March) so that new shoots will emerge below the cut point. The emerged shoots should again be pruned to 50% of their total length after another 3-4 months (May-June). This will in development of proper canopy and initiation of fruiting from the first year of planting.
- After one year of planting fertilizer application should be initiated. For one year old plants, a fertilizer dose of 32 g Urea, 84 g SSP and 32 g MOP should be applied in the month of June and another dose of 32 g urea should be applied in the month of September. In plants of age two year and above, a dose of 64 g urea, 168 g SSP and 64 g MOP should be applied in the month of June and another dose of 64 g urea should be applied in the month of September.
- Pruning operation should be carried out thrice a year during February, May and October. In bearing plants, pruning should be done at the junction point between green tissue and brown tissue in the shoot. In case of fruit bearing shoots, the portion of shoot above the fruits should be pruned.
- The plants are to be severely pruned to secondary branches during 6th year to avoid overcrowding. Fruiting in such pruned plants can again be induced be three times pruning of emerged shoots as described earlier.
- Foliar application of water soluble boron @ 3 ml/litre and Planofix @ 0.6 ml/litre at the time of emergence of new leaves helps in improving fruit quality.
- The plants should be irrigated regularly i.e. once a week during winter season and twice a week during summer season.
- With this method of pruning, fruiting can be obtained in July-August, November-January and March-April.

- Incidence of fruit fly during the month of August is a major concern. Core should be taken not to leave drooped fruits in the orchard ground. Use of methyl euginol trap @ 20 traps per ha during June-July is an effective measure for control of female population of fruit fly. In case of unavailability of the traps in the market, the farmers can contact Head, ICAR RCER, Research Centre, Ranchi, Plandu, Ranchi during the month of May for obtaining the trap.
- Incidence of leaf eating insects can be minimized by spraying Savin @ 1.25 g/ltr at the time of emergence of new leaves. However, spraying should be done after harvesting of fruits.
- For control of termite, chloropyriphus should be used atleast thrice a year.

Economics of High Density Orcharding

Mango

- For establishment of 1 ha of high density orchard of mango cvAmrapali (1600 plants per ha), an amount of Rs. 1.83 lakh will be needed whereas an amount of Rs. 1.00 lakh per ha will be needed for establishment of high density orchard of vigorous cultivars (400 plants per ha).
- For maintenance of 1 ha of the ultra-high density orchard of cv Amrapali (1600 plants per ha), a total of Rs. 3.50 lakh will be required over a period of 10 years whereas an amount of Rs. 2.00 lakh per ha over a period of 10 years towards maintenance of high density orchard of vigorous cultivars (400 plants per ha)
- A total income of Rs. 18.0 lakh can be obtained per ha of high density orchard of cv. Amrapali (1600 no. of plants) over a period of ten years which can amount to a net total profit of Rs. 13.5 lakh per ha in a period of 10 years. In case of high density orcharding of vigorous cultivars (400 plants per ha), a total income of Rs.8.00 will be obtained which can amount to a net profit of Rs. 4.9 lakh in a period of 10 years.
- After 10 years a net profit of Rs. 3.4 lakh can be obtained per ha/year under high density orchard of cv. Amrapali (1600 plants per ha) whereas in case of high density orchard of vigorous cultivars, a net profit of Rs. 4.5 lakh per ha can be obtained in every two years.

Guava

- For establishment of 1 acre of ultra high density orchard of guava, an amount of Rs 1.00 lakh will be needed whereas an amount of Rs. 24000/- will be needed for establishment of 1 acre of traditional guava orchard.
- For maintenance of 1 acre of the ultra high density orchard a total of Rs. 1.77 lakh will be required over a period of 4 years whereas an amount of Rs. 36000/- will be required for maintenance of traditional guava orchard for 4 years.
- A total income of Rs. 2.90 lakh can be obtained per acre over a period of four years which can amount to a net total profit of Rs. 1.16 lakh per acre in a period of 4 years. Under traditional system of guava orcharding a gross amount of Rs. 20000/- can be obtained over a period of four years after establishment.
- After four years a net profit of Rs. 1.06 lakh can be obtained per acre/year under ultra high density orchard whereas a net profit of Rs.24000/- per year can be obtained under traditional guava orchard after four years.

Multitier Cropping Systems

Fruit trees are efficient enough in providing higher economic return even under stressed growing conditions prevailing under the upland situations.The approach aims at improving productivity by effective utilization of air space which is not utilized in single tier system. The multitier system aims at sustainable management of natural resources like soil, water, space and environment. In case of annual crops, 74% of roots do not go beyond 50 cm soil depth whereas in case of perennial crops, top 50 cm almost devoid of feeder roots. 75% feeder roots of perennials remain between 50 and 150 cm depth and 80% active roots confined to 2-3 m radius in grown up trees. The canopy of perennial crops utilizes only 20-30% of land area up to 10 years. Thus, only 25-30% of the land resource is utilized by perennial trees and only 50-100 cm soil depth is utilized by inter crops in uplands which offers ample scope for exploitation of soil depth and inter-space potential through multi-storied cropping The spatially differential root distribution of different component crops in the system helps in higher nutrient and water use efficiency of the multitier system as a whole as a result of presence of feeder roots of the component crops at different depths. Again, the increase in organic carbon content in the soil resulting due to decomposition of fallen leaves from the fruit trees contribute towards enhanced biological activity in the system leading to environmental stability in the rhizosphere. The summation of productivity of the component crops can increase the total productivity of the land where poor soil fertility coupled with low water holding capacity of the soil contributes towards low productivity of any crop in this region. Higher labour requirement per unit area of multitier system contribute towards creation of job opportunities at site. High return per unit area under upland conditions is the ultimate result of fruit based multitier agri-horti system.

The Principles of Fruit Based Multitier System

- Maximization of photosynthetic efficiency of the overall system through efficient reception and utilization of sunlight.
- Phenological compatibility of different component crops helps in fulfilling agronomical requirements of each crop
- Allelopathic interaction between different component crops
- Spatial difference in uptake zone for minimization of competition among the component crops for soil nutrient and water

The fruit based cropping systemconsists of three main components *viz.* main crop, filler crop and inter crop.

Main crop: Main crops are the fruit species having a larger canopy size and prolonged juvenile as well as productive phase. Generally the crops utilize the entire land after 20-25 years whereas only 25-30 per cent of land is effectively used up by the main crop up to 10 years. These plants are planted at wider spacing.

Criteria for selecting of main crops

1. Xeric characters such as deep tap root system, high bound water in tissues, sunken stomata, waxy coating, thick cuticle and pubescence.
2. Ability to complete maximum vegetative and reproductive phase during the period of moisture availability.
3. Root system and root growth should be able to exploit deeper soil layers than those tapped by the under and ground storey crops.

4. Branching habit that allows light penetration to the under storey crops.
5. Species should be strong coppicer and should respond to pruning.
6. Leaf fall during the growth period of the ground crops and the rate of litter fall and decomposition should have positive effect on soil fertility.
7. Resistant to pest and diseases.

Filler crops: The filler crops are the fruit species which are precocious in nature, prolific bearer having short stature. These plants are planted with the purpose to generate additional income from the land during the juvenile and initial bearing stage of the main crops. The plants generally hardy in nature and have shorter economic life than main crop and planted within the main crop at a closer spacing. The filler plants can be removed after the main crops attain effective canopy size for yielding economically.

Guava, being a prolific and precious bearing fruit plant and dwarf stature canopy can very well be fitted as second storey crop for additional income at least in the initial years of the orchard (Nath *et al.*, 1997).

Table 6.1 Suitable varieties of different fruits crops

Mango	Bombay Green, Himsagar, Zardalu, Krishna Bhog, Langra, Maldah, Dashehari, Chausa, Mallika, Amrapali
Litchi	Shahi, Ajhauli, Rose Scented, Trikolia, Swarna Roopa, China
Aonla	Narendra Aonla-7, Kanchan
Sapota	PKM-1, Kalipatti, DHS-1, Cricket Ball
Jack fruit	Khajva, Swarna Poorti, Swarna Manohar Guava Sardar, Allahabad Safeda, Arka Mridula
Custard apple	Balanagar, Arka Sahan
Papaya	Pusa Dwarf, Pusa Nanha
Lime	Kagzi

*Inter crop:*The inter crop occupy the lower most layer of the system and are grown in the remaining unused land. Generally the inter crops are the location specific annual crops, selected as per the climatic and socio-economic suitability. During the initial years any crops can be taken whereas at later stage shade tolerant crops can be grown as inter crops.

Criteria for selecting of inter crops

1. Early maturing.
2. Should have shade tolerance suiting the amount of available solar radiation.
3. Should not compete with the perennial companion crop.
4. Should not be more susceptible than the main crop to diseases they have common.
5. Should not require mechanized harvesting or other operations that would damage the main crop or induce soil erosion or damage soil structure.
6. Should not have economic life longer than that of main crop.
7. Should suit the local agro climatic, labour, irrigation and market/ processing facilities.
8. Leguminous ground storey crops that have nitrogen fixing capacity should be preferred.

Leguminous crops have been reported be the most remunerative intercrops in fruit based cropping system during the initial 7-10 years (Nath*et al.*, 2004; Bhatnagar and Joshi, 1998). Bhatnagar and Joshi (1998) also mentioned improvement in soil environment and fertility due to intercropping of legumes in fruit based cropping system in central, southern and northern plain zones. Sharma and Vashishtha (1985) have also advocated monocot grass spp. (*Cenchrus ciliaris*) under stress conditions of arid region. Many tropical MAPs are well adapted to partial shading, moist soil, high relative humidity and mild temperatures (Vyas and Nein 1999), allowing them to be intercropped with timber and fuel wood plantations, fruit trees and plantation crops. Some well known medicinal plants that have been successfully intercropped in India, include *safedmusli*(*Chlorophytum borivilianum*), rauvolfia (*Rauvolfiaserpentina*), turmeric (*Curcuma longa*), wild turmeric (*C. aromatica*), *Curculigo orchioides*, and ginger (*Zingiber officinale*) (Chadhar and Sharma 1996; Mishra and Pandey 1998; Prajapati *et al.*, 2003).

Allelopathy effect in fruit based cropping system

The term allelopathy, was introduced by Molisch in 1937, and is derived from the Greek words allelon 'of each other' and pathos 'to suffer' and mean the injurious effect of one upon the other. In 1996 The International Allelopathy Society defined allelopathy as follows: *"Any process involving secondary metabolites produced by plants, micro-organisms, viruses, and fungi that influence the growth and development of agricultural and biological systems (excluding animals), including positive and negative effects"* (Torres *et al.*, 1996). Chemicals released from plants and imposing allelopathic influences are termed allelochemicals or allelochemics. Allelochemicals can be present in several parts of plants including roots, rhizomes, leaves, stems, pollen, seeds and flowers. Generally, there are four processes by which allelopathic substances are released into the environment: volatilization, leaching, decomposition of plant residues, and root exudation. When susceptible plants are exposed to allelochemicals, germination, growth and development may be affected. Allelopathy occurring among individuals of the same species is termed autotoxicity. Autotoxicity is known for example in *Medicago sativa* (alfalfa), *Trifolium* spp. (clovers) and *Asparagus officinalis* (Asparagus).

A number of crops and vegetables are known for their allelopathic activities such as corn (Almezori, 1996), sunflower (Schon and Einhellig, 1982), sorghum (Alsaadawi *et al.*, 1993), and tomato (Saied and Saied, 2001). Tomato seed germination, seedling growth and dry weight are very sensitive to aqueous extract of most tested plants [okra (*Hibiscus esulentus*L.); sorghum (*Sorghum bicolor* L. subspecies *tichnicum*); sunflower (*Helianthus annuus*L.); pepper (*Capsicum frutescens*L); and tomato (*Lycopersicon esculentum* L)]. This means that compatibility will not occur between tomato and these interacted plants in mixed farming. At the same time okra plants are less susceptible to the aqueous extracts of all tested plants except that of sorghum. This means that okra exhibits more resistance toward phytotoxic substances from other intercropped plants and can grow together at the same time. (Ali and Mezori, 2008)

Although black walnut is considered as the most notorious of allelopathic trees (Rietveld, 1983),it has long been recognized that the principal chemical responsible for walnut allelopathy is a phenolic compound called Juglone (5-hydroxy-1,4- naphthoquinone) (Davis, 1928). However, installation of polyethylene root barriers proved to be efficient in preventing juglone from getting into the alley where associated crop species are normally planted. This implies that management practices such as root pruning, fertilizer injection, or root discing can be used to limit the impacts of juglone.

Performance of fruit based multitier agri-horti systems under upland

In a trial conducted under Ranchi condition, *Mango based multitier agri-horti system:*different intercrops differed with respect to their effect on growth and yield of mango as well as different filler plants. With respect to plant growth of mango plants, intercropping of paddy resulted in the maximum plant height (4.63 m) whereas the minimum tree height was recorded in case of fallow (3.45 m) (Fig. 6.1A). As evident from the figure 1, the filler plants did not have any effect on plant height of mango. With respect to canopy spread, the filler crops did not have any effect. However, intercropping of paddy resulted in the maximum canopy spread of mango (4.85 m) and the minimum canopy spread was recorded with fallow (3.31 m) (Fig. 6.1B). Similarly, with respect to trunk cross-sectional area, the maximum value was recorded in case of paddy as intercrop (427.43 cm2) whereas the minimum value was recorded in case of fallow (171.08 cm2) (Fig. 6.1C). It is clear that enhanced growth of mango plants with paddy as intercrop. However, with respect to cumulative yield per tree during 2005-10, the minimum value was recorded with cowpea as intercrop whereas the yield of mango did not differ markedly with other intercrops (Fig. 6.1D).

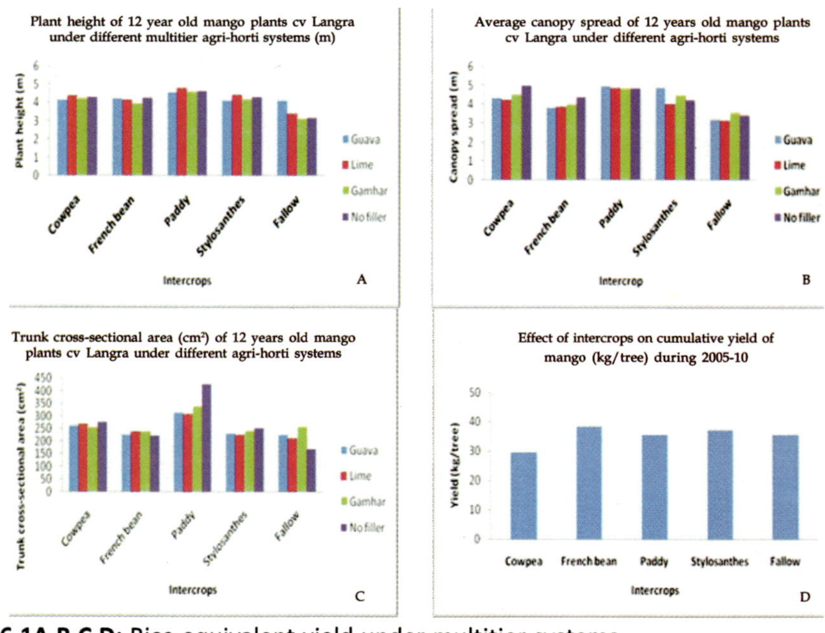

Fig 6.1A,B,C,D: Rice equivalent yield under multitier systems

Table 6.2 Rice equivalent yield (t/ha) under different mango based multitier agri-horti system

Mango based cropping systems	Rice equivalent yield (t/ha)					
	2005-06	2006-07	2007-08	2008-09	2009-10	Cumulative
Mango+Guava+ Cowpea	8.70	3.99	7.05	7.14	7.40	34.28
Mango+Citrus+ Cowpea	13.92	5.57	2.53	2.47	2.35	26.84
Mango+Gamhar+ Cowpea	0.09	0	2.79	2.99	2.01	7.89
Mango+nofiller+ Cowpea	0.19	0.96	2.00	1.84	2.25	7.24
Mango+Guava+ Paddy	10.92	8.75	6.88	8.40	8.61	43.56
Mango+Citrus+ Paddy	8.26	0.74	1.14	0.98	2.11	13.22
Mango+Gamhar+ Paddy	0.53	0.64	1.76	1.23	2.00	6.16
Mango+nofiller+ Paddy	0.32	0.48	1.63	0.30	2.62	5.35
Mango+Guava+ Stylosanthes	11.11	6.04	7.77	10.60	8.11	43.63
Mango+Citrus+ Stylosanthes	4.12	4.72	0.57	2.11	2.65	14.18
Mango+Gamhar+ Stylosanthes	0.01	0.00	1.32	0.41	1.91	3.64
Mango+nofiller+ Stylosanthes	0.00	0.00	0.90	0.41	1.62	2.94
Mango+Guava+ French bean	14.69	10.14	8.73	9.52	8.16	51.24
Mango+Citrus+ French bean	8.37	2.40	2.76	6.23	2.56	22.32
Mango+Gamhar+ French bean	4.33	0.96	2.64	2.76	2.11	12.80
Mango+nofiller+ French bean	3.92	0.29	3.02	0.59	2.30	10.11
Mango+Guava+ Fallow	11.44	5.12	4.78	7.12	7.23	35.69
Mango+Citrus+ Fallow	6.43	1.06	0.90	3.07	2.39	13.84
Mango+Gamhar+ Fallow	0.00	0.00	0.74	1.76	2.16	4.66
Mango+nofiller+ Fallow	0.00	0.00	0.42	0.64	2.00	3.06

Among the filler plants, the maximum height of guava was recorded with French bean as intercrop whereas the maximum values of trunk cross-sectional area was recorded with paddy as intercrop. With respect to cumulative yield per tree, the maximum value was recorded with intercropping of Stylosantheshamata and intercropping of cowpea resulted in the minimum yield of guava.

Aonla based multi-storeyed cropping system: Aonla has sparse foliage which allows 87.5% area for intercropping during the initial 10 years. The experiment was initiated to standardize suitable intercrops in aonla based multitier system with guava as filler crop during the pre-bearing stage of main crop grown under rainfed conditions. During initial years, no adverse effect of inter crops on main and filler crops and vice versa was observed. After nine years of planting, none of the treatments differed significantly with respect to the plant growth parameter of main and fill crops.

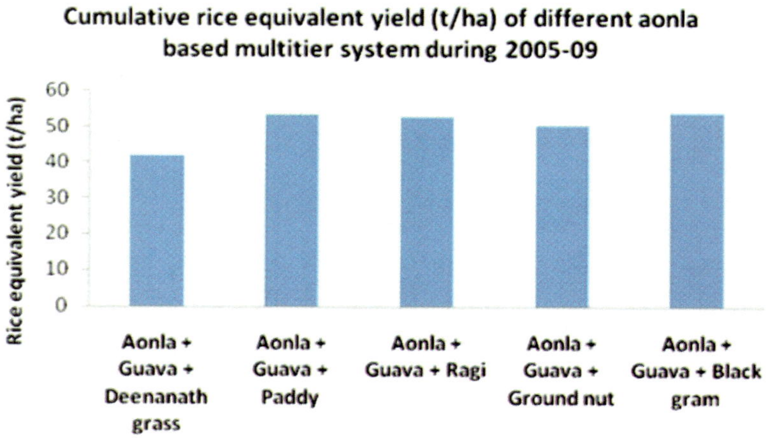

Fig 6.2 Performance of aonla based multitier system

*Litchi based multitier cropping system:*Under plateau region of eastern India, litchi plants have very slow rate of growth during initial phase of orchard establishment. Litchi based agri-horti systems can provide a profitable alternative under the plateau conditions. Keeping this in view an experiment was initiated under the National Agricultural Technology Project on "Development of fruit based land use system in watershed" with the objective to standardize suitable intercrops under litchi based multitier system with guava as filler crop during the pre-bearing stage of main crop. Litchi plants. Different intercrops like Cowpea, French bean, niger and horse gram were grown in the interspaces. Intercropping did not show any marked adverse effect on mean monthly increment in the height of plant, girth of trunk and area of canopy spread of main crop as well as filler crop. Fruit yield of filler crop was (guava) also not influenced by intercropping. Among the intercrops, cowpea was found to be the most profitable (Rs 24,000/ha). After six years, drastic reduction in the yield of all the intercrops could be observed due to the shading effect of main and filler crops. At the end of 8th year, the cowpea continued to record its supremacy as an intercrop through performing poorly, over other intercrops and yielded 3.03 t/ha green pods. Keeping this in view, intercropping of cowpea was recommended under the litchi based multitier cropping system upto the initial 8 years of the orchard. Afterwards, intercropping of shade tolerant crops like ginger, turmeric and elephant foot yam can be followed.

7

Vegetable Based Production Systems

The eastern region of India covers states like Bihar, Jharkhand, eastern Uttar Pradesh, Odisha, West Bengal, Chhattisgarh and Assam. It covers 28% of the national geographical area and 35% of total population of country. The population density is 1.24 times to that of national average. The region is blessed with adequate rainfall for successful cultivation of different horticultural crops with rainfall ranging between 1200 mm to 2400 mm (West Bengal: 1750 mm, Bihar: 1204 mm, Orissa: 1450 mm, Assam: 2477 mm and Jharkhand: 1300mm). With respect to vegetable production, the region contributes nearly 44% of total production of India from 45.81% of total area under vegetable cultivation in India. This indicates the importance of eastern India in total vegetable production in the country.

The climate of the eastern region is tropical, hot and humid except in hilly areas with high rainfall. The eastern plains are endowed with rich basic natural resources viz., the most fertile land and abundance of water. The average annual rainfall in this region varies from 1008 mm to 3126 mm. The mean annual rainfall over the geographical domain of eastern UP, Bihar plateau, Bihar plains, Orissa, sub-Himalayan West Bengal, and Gangetic West Bengal is about 1008 mm, 1373 mm, 1203 mm, 1482 mm, 3126 mm and 1425 mm, respectively. Even though the region has abundant rainfall, surface and ground water resources, they are grossly underutilized.

Large proportion of the cultivated area does not receive irrigation and the farmers depend on the monsoon for crop production. In spite of the fact that crops in rabi season are relatively disease free and receive plenty of sunshine, large tracts of cultivable land during rabi season remain fallow. As sizeable part of the cultivated area in eastern region do not have provision for assured irrigation. Therefore, even short spell drought adversely affects the stability of agricultural production, thereby resulting in low productivity. As a result, the employment in agriculture sector is limited and a large proportion of the population still remains below the poverty line and suffers from malnutrition. The region is rich in rain, surface and ground water resources with plenty of sunshine and large tract of fertile land. However, owing to poor utilization of water resources, the cropping intensity in the region is low (142%).

Horticulture and horticulture based cropping patterns offer best land utilization option for these areas. Cereals occupy the major share of agricultural production in eastern India. A look at the productivity of different cereal crops indicates lower productivity in all the states except West Bengal and Bihar (Table 7.1).

Table 7.1 Productivity of cereals (t/ha)

States	Rice	Maize	Wheat
Assam	1.33	0.73	1.13
Bihar	1.48	2.74	2.00
Jharkhand	1.20	1.40	1.70
Odisha	1.55	1.50	1.40
West Bengal	4.00	3.40	2.60

At present, the area under irrigation at national level is 32.9%. Among the eastern states, only Bihar (45.31%) has more area under irrigation than that at national level (Table 7.2). Hence, rainfed agriculture contribute a major share of agricultural production in the region.

Table 7.2 Area under irrigation

States	% irrigated area
West Bengal	32.65
Odisha	23.94
Bihar	45.31
Jharkhand	11.08
Chhattisgarh	23.79
All India	32.91

With respect to nutrient use in agriculture production, most of the eastern states rank much lower than the national average. Nutrient use in Bihar and West Bengal are higher than that at national level (Table 7.3). So there exists a large scope for increasing the productivity of different crops through increasing nutrient use in the eastern region.

Table 7.3 Nutrient use (Kg/ha)

States	N	P_2O_5	K_2O	Total
Assam	30.3	12.5	15.0	57.8
Bihar	116.8	31.5	20.6	168.9
Chhattisgarh	30.7	11.9	3.9	46.5
Jharkhand	36.0	19.0	5.0	60.0
Odisha	34.3	17.0	10.3	61.6
West Bengal	71.6	42.6	41.6	155.8
All India	59.2	22.1	8.5	89.8

With respect to area under vegetable crops (Fig 7.1), the percentage increase in area as well as production from 2001-02 to 2012-13 has been significant both at national level and eastern India level. The increase in area at national level was much higher than that recorded under eastern India conditions. However the eastern region has witnessed marked increase in production during 2001-02 to 2012-13 as compared to that at national level.

As evident from Fig 7.2, the maximum increase in area under vegetables was recorded in states like Chhattisgarh, Jharkhand and Bihar. With respect to vegetable production, the significant increase recorded in eastern region was attributed to contribution from Chhattisgarh, Jharkhand, Bihar. A look at the changing productivity of vegetables indicates marked increase both at national level as well as eastern India level and as evident in Fig 7.4, the percentage increase was similar at national as well as eastern India level. However, during both the years the vegetable productivity of eastern India was lower than that at national level. This warrants for concerted efforts for increasing the vegetable productivity in eastern India.

A look at changing productivity of vegetables in different state indicates increase in productivity in states like Jharkhand, Odisha, eastern UP and West Bengal (Fig 7.5). The increase vegetable productivity in Bihar was most remarkable. However, the stagnant productivity in Assam and Chhattisgarh is a matter of concern.

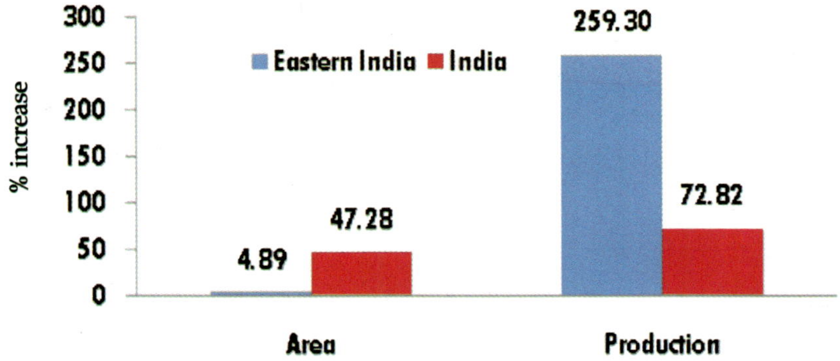

Fig 7.1 Percentage increase in area and production of vegetables during 2001-02 to 2012-13

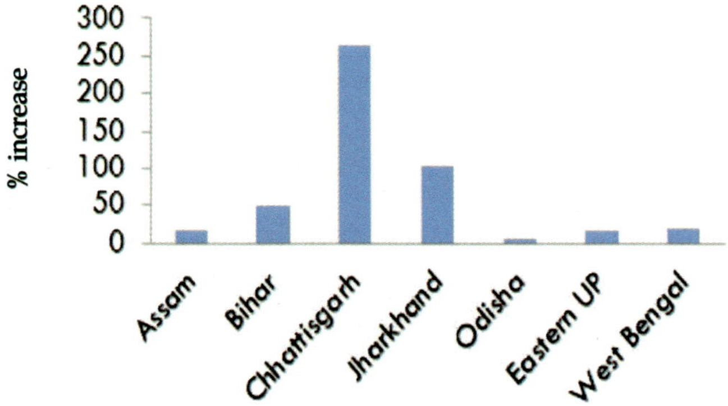

Fig 7.2 Percentage increase in area under vegetable crops during 2001-02 to 2012-13

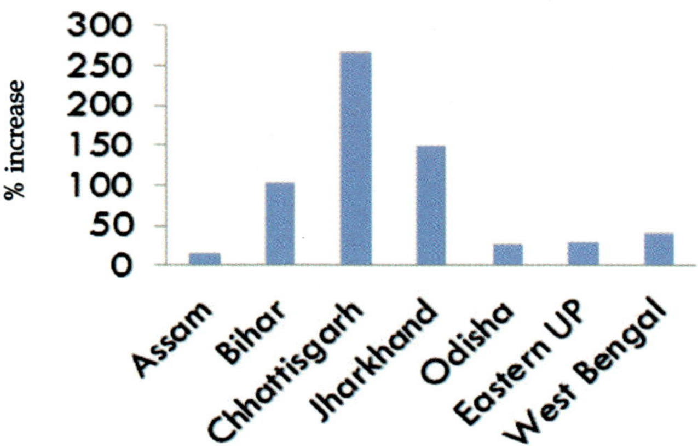

Fig 7.3 Percentage increase in production of vegetable crops during 2001-12 to 2012-13

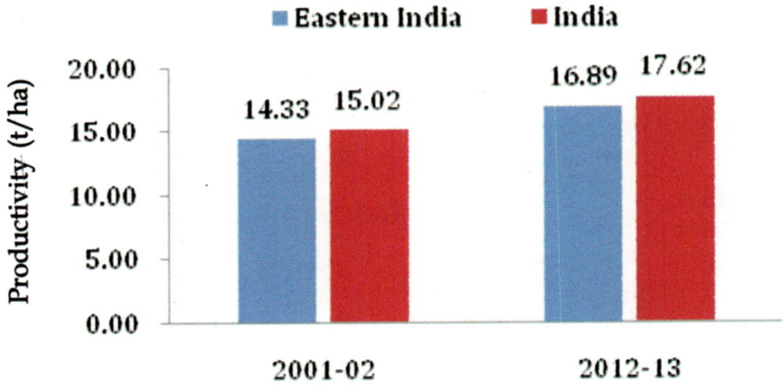

Fig. 7.4 Changing productivity of vegetable crops

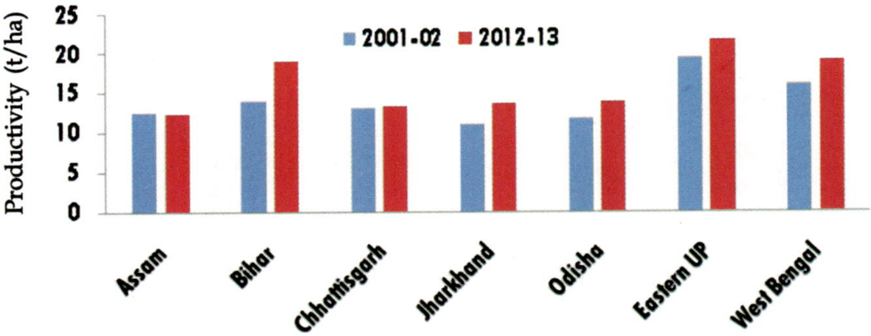

Fig. 7.5 Productivity of vegetables in different eastern states

A look at changing pattern of area under different vegetables in eastern India indicates marked decline in share of area under sweet potato and significant increase in area under potato. However, at national level, the contribution of area under potato increased by only 2%. It is also noteworthy that the share of area under onion increased markedly at national level as compared to that in eastern India. Similar trend was also recorded with respect to vegetable production in eastern India and India.

Productivity of different vegetables like brinjal, cabbage, cauliflower, okra, onion, peas under eastern Indian conditions was very high during 2001-02 as compared to productivity levels at national level (Fig 7.6).

During 2012-13, although the productivity of all the vegetables increased, but the productivity gap between eastern India and national level narrowed down significantly indicating Eastern India has to play a major role in fulfilling vegetable requirement of the country.This can be attributed to large-scale introduction of improved varieties of different vegetable crops along with high input based production technologies by the farmers.

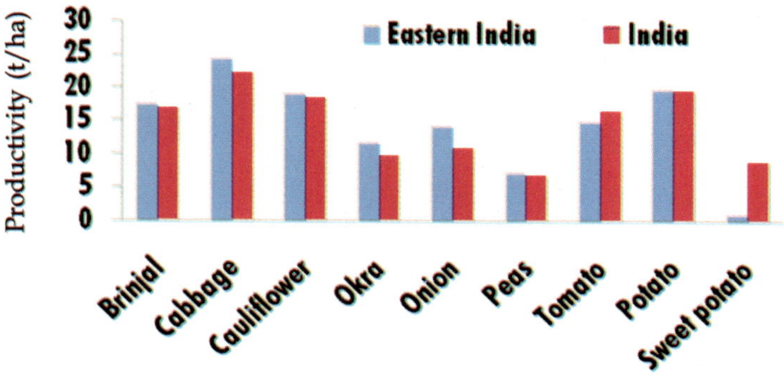

Fig 7.6 Productivity of different vegetables during 2001-02

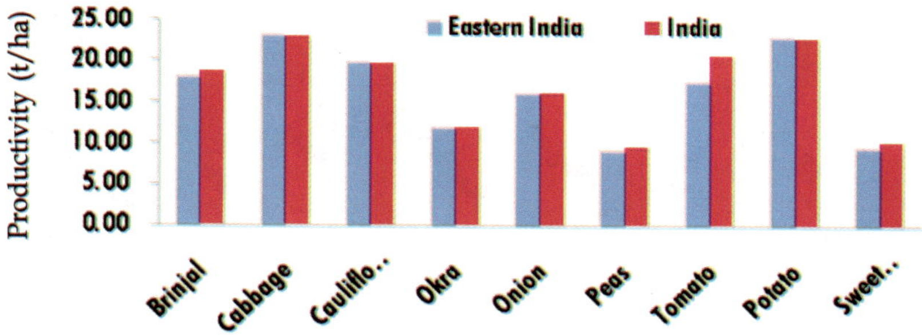

Fig 7.7 Productivity of different vegetables during 2012-13

Table 7.4 Surplus availability and requirement of vegetables in eastern India

	Requirement (MT)	Availability (MT)	Requirement (MT)	Projected Availability (MT)
	2013		2026	
Eastern India	35.68	56.99	49.64	81.8
India	133.98	129.75	153.28	140.83

There is a need for increasing the profitability of vegetable based cropping systems to make it attractive for the farming communities of eastern India. Work done under Jharkhand conditions indicates high level of dividends under different vegetable based crop rotations which makes it an attractive venture (Fig 7.8). The outcome of studies conducted on crop rotations being followed by the vegetable growers around Ranchi are detailed in Table 7.5.

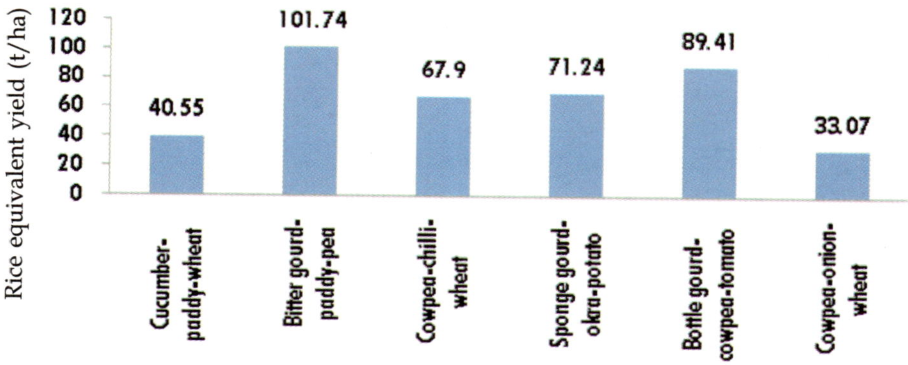

Fig. 7.8 Vegetable based cropping system: Profitable under uplands of eastern region

The ICAR-RCER, Research Centre, Ranchi has been a pioneer for development of technologies for in creaning productivity of vegetable in concern region some of the technological options for increasing productivity of vegetables are listed below:

1. Improved varieties in vegetable crops

A large number of varieties and hybrids have been identified and developed for commercial cultivation in brinjal, tomato, pea, cowpea, cucumber, french bean, dolichos bean, winged bean, vegetable soybean, parwal, kundru (*Coccinia grandis*), ridge gourd, sponge gourd, kakrol (*Momordica cochinchinensis*). Benefit cost ratio of improved varieties is detailed in Table 7.6. Bacterial wilt is the major disease of tomato and brinjal and we have already released bacterial wilt resistant varieties of tomato and brinjal.

2. Potential Underutilized leafy vegetables

Amaranthus sp, *Ipomea aquatic, Basella alba, Chenopodium album, Antidesma diandrum, Cassia tora, Vangueria spinosa, Brassica rugosa,* Chinese cabbage, Drumstick etc. play an important role in achieving nutritional security.

3. Potential tuber crops

Germplasm collected from Jharkhand, Chhattisgarh, Odisha and West Bengal viz., *Dioscorea* spp. (47), *Ipomoea batatas* (12), *Maranta arundinacea* (1), *Colocasia* spp. (67), *Amorphophallus* spp. (47), *Alocasia*spp. (15), *Manihot* spp. (7) have been multiplied in the field for characterization and evaluation.

Table 7.5 Paddy equivalent yield realized in different vegetable based crop rotation in Ranchi

Block	Crop rotation	Paddy-equivalent yield realized (q/ha)
Bundu	1- Paddy-cabbage-sponge gourd	160.00
	2- Paddy-tomato	152.00
	3- Brinjal-sponge gourd	180.00
	4- Paddy-potato	170.00
	5- Paddy-tomato-chilli	232.00
	6- Brinjal	120.00
	7-Cauliflower-onion-sponge gourd	280.00
	8- Paddy-cauliflower+spinach	140.00
Tamar	1- Paddy-gram	120.00
	2- Brinjal	120.00
	3- Cucumber+dolichos bean	140.00
	4- Tomato	132.00
	5- Okra-paddy-okra	120.00
Silli	1- Paddy-wheat-okra	112.00
	2- Brinjal-potato	170.00
	3- Tomato-potato-bitter gourd	342.00
	4- Cucumber-maize-paddy	102.00
	5- Cucumber-cauliflower	140.00
Lapung	1- Paddy-wheat-bitter gourd	122.00
	2- Paddy-potato-pumpkin	230.00
	3- Sweet potato-pea	167.50
	4- Paddy-pea	100.00
	5- Paddy-wheat	62.00
	6- Paddy-chilli	100.00
	7- Tomato	132.00
Ormanjhi	1- Paddy-bitter gourd	80.00
	2- Brinjal-cauliflower+ginger	260.00
	3- Tomato-cauliflower+ginger-bitter gourd	352.00
	4- Cucumber-potato	190.00
	5- Chilli-potato-onion	350.00
	6- Onion-cauliflower	120.00
	7- Sweet potato	87.50
	8- Brinjal	120.00
	9- Tomato	132.00
	10-Maize+cowpea+ridge gourd-potato-onion	329.400

Table 7.6 Benefit cost ratio of vegetable varieties

Technology	Benefit cost ratio
Biotic stress resistant varieties of brinjal (Swarna Pratibha & Swarna Shyamli)	2.99:1
High yielding hybrids of brinjal (Swarna Shakti and Swarna Ajay)	3.05:1
Biotic stress resistant varieties of tomato (Swarna Lalima& Swarna Naveen)	2.85-2.95:1
High yielding hybrids of tomato (Swarna Samridhi, Swarna Sampada, SwarnaVaibhav& SwarnaVijaya)	2.71-3.16:1
Improved varieties of pointed gourd (Swarna Rekha, Swarna Alaukik & Swarna Suruchi)	1.90:1

4. Drip irrigation

Placement of laterals at 10 cm soil depth resulted in highest water use efficiency of bitter gourd (6.8 kg/m^3). Water productivity up to 412.7 Rs/m^3 recorded in tomato through fertigation.

5. Resource conservation techniques

Efficacy of different methods of planting and soil moisture conservation techniques in cauliflower-brinjal-tomato crop sequence can be used to improve the land and water productivity. In cauliflower, Ridge and furrow system with plastic mulch was found better than flat bed and broad bed method (Table 7.7). Use of plastic not only resulted in increase in yield of cauliflower but also increase in water productivity.

6. Plastic mulching with drip irrigation

Drip irrigation with plastic mulching provides optimal use of water and nutrients which maximizes the harvest.

Table 7.7 Performance of Ridge and furrow system-Cauliflower

Treatment	Yield, t/ha	Water Productivity, kg/m^3	Economic water Productivity, Rs/m^3
Without plastic	22.5	9.33	93.58
With plastic	29.4	12.55	122.5

7. Inter generic grafting for bacterial wilt resistance

Susceptibility of high yielding F_1 hybrids of solanaceous vegetables like tomato, brinjal, capsicum to bacterial wilt prevent their cultivation during rainy season. The available bacterial wilt resistant varieties of tomato and brinjal can be used as rootstock to prevent mortality during rainy season. Tomato hybrid SwarnaVaibhav grafted on bacterial wilt resistant brinjal rootstock HAB-900 and HAB 901 recorded maximum percentage of plant survival against wilt (96%).

Fig. 7.9 Tomato grafted on brinjal

Fig 7.10 Raising of nursery in protray

8. Protray nursery raising of seedlings

Protray nursery raising provides the advantage of healthy seedlings in less time (around two weeks).

9. Protected cultivation of vegetables

Vegetable production under protected structures is gaining importance in south East Asia. Tomato is the most universally grown greenhouse vegetable crop found in varied agro-climatic conditions. Indeterminate varieties that set fruit for almost a year are economically profitable under protected cultivation. Swarna Anmol. (180t/ha) and Swarna Kanchan, variety for processing (50-55t/ha); with high yielding capacity and multiple disease resistances (bacterial wilt, TMV & F_1) have been developed for cultivation in polyhouse conditions.

Fig 7.11 Naturally ventilated Polyhouse

Fig 7. 12 Tomato variety SwarnaKanchan for protected cultivation

Fig 7.13 Capsicum under protected cultivation

Fig 7.14 Cabbage under protected cultivation

Future strategies for vegetable sector in eastern India

There must be increased emphasis on post-harvest management to reduce the supply demand gap and provide a consistent income to the farmers. Crop improvement for biotic and abiotic stress resistance is a pre requisite to decrease the cost of cultivation. Optimization of inputs should be achieved through increasing water and nutrient use efficiency of vegetable based production system. Increase productivity of underutilized vegetables for providing nutritional security. A sound seed production programme for making available quality seedsof different vegetables to the farmers.Vegetable crops will continue to play a significant role in the agrarian economy of eastern India. Eastern India has to play a major role to fulfill the vegetable requirement of India. Investment in post-harvest management will go a long way in strengthening the vegetable production economy of the region.

□□□

8

Water Productivity of Livestock and Fisheries

A country is considered as water stressed if the annual renewable fresh water resources are below 1,700 m^3per capita per year and the water scarcity threshold would be below 1,000 m^3per capita per year. The number of water scarce countries is expected to increase from 20 in 1990 to 35 by 2020. Our country will also fall under this category. We are already experiencing water stress and scarcity in many regions of the country. In agricultural economy of India, livestock plays a major role in mixed crop-livestock system and water need of livestock cannot be ignored.

Livestock contribute about 27% of India's agricultural GDP, sustain millions of pastoral families, and provide income for about 80% smallholder farmers in India beside contribution for draft power & manure and serve as the asset base for hundreds of millions of poor farmers mostly in the water stressed areas having no access to financial institutions.

Due to change in food habit, demand for livestock products has risen strongly in the last decades and is expected to increase further particularly in India and other developing countries, thus augmenting pressure on already scarce resources and increasing the necessity to improve water productivity in sustainable ways (Table 1). Livestock water demand for direct consumption is expected to increase by 71% worldwide by 2025, the highest increase is to be taking place in the developing countries like India. In recent years, livestock production has received widespread criticism due to the use of grain for feed resources because large amount of water is needed to grow these feed resources.

Table 8.1 Required growth rate of livestock products in India (moderate income growth)

Products	Growth rate (%)		
	1993-2020	1993-2000	2000-2020
Milk	4.88	4.48	4.82
Mutton & goat meat	13.68	9.46	13.62
Beef & buffalo meat	3.45	3.32	3.39
Chicken	4.77	4.60	4.72
Egg	6.18	6.02	6.11

(Source: Dastigiri, 2004)

Considering the increasing water scarcity in many areas in the world and the expected increasing demand for livestock products, the need to understand, define and increase the water productivity of livestock related production systems is evident.

Why livestock and water?

The water required to produce food for people is about 3500 litre/day/adult for a 3000 Kcal diet, or about 70 times more than that needed for other domestic purposes including safe drinking water as per a report in Nile basin. The population in the country will double over the next 20 years, creating greater demand for water to produce food unless improved agricultural water productivity options are available and used. Huge livestock population in the country needs water for drinking and producing feed. Production of livestock feed requires as much water as human food crops. In the past, water productivity concepts have been applied mainly to crop production. But as demand for milk and meat will likely to double over the next 20 years, livestock production needs to be factored into future assessments of water use and productivity.

What is livestock water productivity?

Livestock water productivity in agricultural systems is the ratio of the sum of animal products and services produced to the amount of water depleted in producing them (Peden *et. al.*, 2006b). Livestock water productivity assessments can use quantitative indicators of animal outputs (e.g., kg of meat, milk and manure or draft power) or indicators of economic and social benefits that people derive from animal keeping. Water accounting assesses the amount of water entering and leaving an agricultural system. It helps to identify options for increasing water productivity involving increased animal production and services and decreased water depletion. This concept is applicable to individual, communities/ watersheds and river basins. The challenge is to manage water, land and animals in ways that use available water to increase transpiration needed for production and environmental health and to decrease evaporation and harmful discharge. Major sources of water depletion include:

- Evaporation from land, water and vegetation surfaces.
- Transpiration from green plants, which drives photosynthesis and is essential for agricultural production.
- Discharge or physical loss of water from agricultural systems.

We have defined livestock-water productivity as *the amount of water depleted or diverted to produce livestock and livestock products and services, including energy*. Often water related to livestock production is assumed to include only water either directly consumed by the animals or used for cleaning and other service functions during the processing of animal products. Development and governmental strategies often consider only watering needs, environmental degradation and pollution of water sources by livestock. The water needed by the animals for drinking is often a major factor limiting production and has to be available in sufficient quantity and quality, although the amount of water needed to produce the daily feed resources can be over 100 times higher than the water consumed.

Drinking water versus water for feed

Livestock water demand is a key component of the agricultural water nexus in crop-livestock mixed farming system prevalent in India though it was neglected both by the researcher and policy makers. Adequate water is required to improve the productivity of both crop and livestock enterprises, but until recently no systemic approach has been made to assess the inter-relationship between crop production and livestock farming in the context of the growing competition for land and water resources. In past, only drinking water need of livestock was considered which overestimated the value of water productivity as 300 liters of water to produce 1 kg meat in Africa. But actual water requirement for livestock is much higher than the drinking water as drinking water requirement is less than 2% of total water requirement.

Recent estimate showed that to produce 1 kg of meat about 100000 liters water was needed in Africa (IWMI Brief Note on Maximizing Benefits from livestock-water venture in Sub-Saharan Africa). This amount of water includes water for drinking and for producing animal feed. Since, drinking water requirement is negligible in comparison to water requirement for producing feed; water requirement for producing feed is a major factor of concern. It is estimated that 1 Tropical Livestock Unit requires 5 kg of dry matter per day and about 1.5 m^3 of water is required to produce this dry matter (Peden *et al*, 2006b). Water productivity of animal feed varies in different zones of IGP depending on different agro-climatic conditions, variety used, management practices and method of estimation etc Water productivity of rice and wheat grain in different regions of IGP varied from 0.46-0.87 and 0.55-1.27, respectively. Similarly, water productivity of maize grain, oilseed and pulses varied from 0.92-1.74, 0.30-2.89 and 0.39-4.95, respectively in IGP. In IGFRI, Jhansi, water requirement to produce 1 kg DM was estimated at 454.5 liters for berseem, 312.5 liters for oat, 267 liters for Sorghum and 555 liters for cowpea (Baig*et a*l, 2007). Preliminary study in Patna showed that water productivity (Rs/ m^3) of wheat and rice is 0.9 to 1.0 and 0.5 to 0.55, respectively. However, such information is incomplete unless factor productivity from horticulture, livestock, fisheries are considered. Water productivity of canal and tube-well command in Bihar was reported at 2.67-3.96 and 3.09-3.68 Rs/m^3 (Sikka *et al.*, 2008). In India, the water requirement for producing 1 kg meat or milk may be lower as animals are mainly reared on crop residues and natural pasture. It is also a fact that major share of water used in crop-livestock system is contributed by green water. In a study in Gujrat, it was reported that agronomic efficiency of water used in milk production was 0.31 liters per m^3 of water for buffalo and 0.49-liters/ m^3 of water for crossbred cows. The economic efficiency of water use was reported to be 0.19 Rs/m^3 for buffalo and 0.17 Rs/m^3 for crossbred cows. The water requirement for 1 kg of milk production was found lower in buffalo (0.31 liters) than crossbred cows (0.49 liters) in Gujarat, since buffalo is better converter of low quality roughage into energy (Singh *et al.*, 2004).

About 450 m^3 of water is required annually to produce feed to maintain one TLU. When animals are growing, working, stressed or lactating, they use even more. Transpired water depleted to produce feed for animal maintenance and production can be 50 to 100 times more than what animals drink (Peden *et al.*, 2006b). Although this relatively small amount of quality drinking water is essential for animal health and production, drinking water is not part of water productivity because water consumed by an animal has not been depleted or lost from the agricultural system. Animal feed sourcing strategies that utilize rangelands and crop residues and byproducts that are unsuitable for human food production can increase water productivity as long as environmental needs are met. Improving animal grazing and watering practices to increase infiltration and reduce run-off and contamination can help improve water productivity. Transpiration is water depletion that enables crop, livestock and tree production and sustains biodiversity and ecosystem services. Water depletion such as flooding and evaporation contribute little value to agriculture and can damage the environment. Water depleted for livestock services such as draught power used for ploughing or transport of crop products to the markets should not be accounted for on the livestock side, but within crop water productivity because this is part of crop production application and to suggest strategies for improving livestock water productivity. The water accounting framework helps analyzing different livestock production scenarios regarding their water productivity, sustainability and environmental impact (Fig. 8.1).

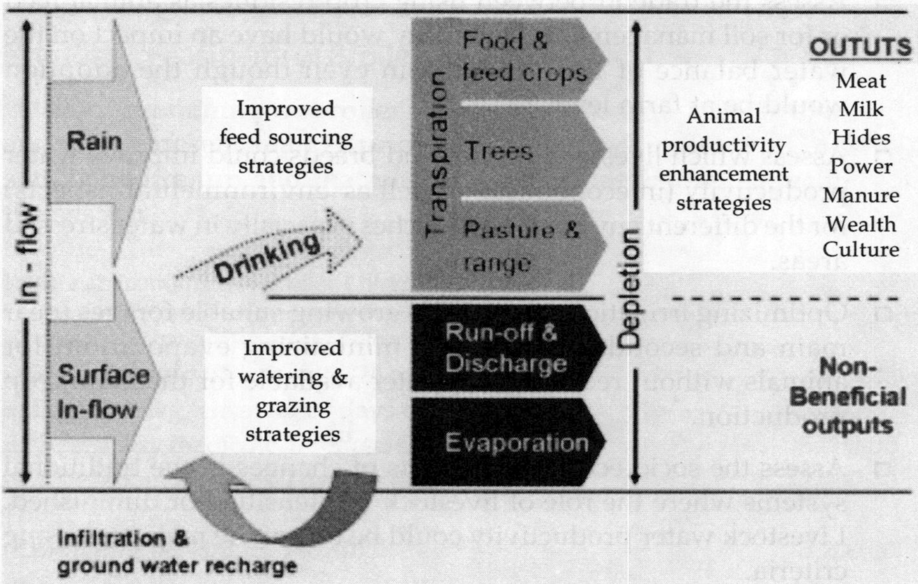

Fig. 8.1 Simplified livestock water productivity framework (Peden *et al.*, 2006)

Strategies to increase livestock water productivity

- Livestock water productivity could be improved through (1) the production of feed sources that would utilize transpired water efficiently and (2) the use of livestock species and breeds that would have higher conversion rates for the available feed. In all the cases, it will be important to consider the sustainability of the agro-ecosystems. Improving the grazing and watering management of livestock will benefit the productive capacity of the land through better soil and water conservation and diminish pollution with positive environmental impacts.
- The livestock water productivity should be quantified at household, watershed/community and basin levels. This would require developing a methodology for including livestock water productivity into the water productivity equation.
- Improving the quantity and quality of crop residue especially the protein through the introduction of legume forages without affecting the major crop should be looked into from the socio-economic perspective.
- Assess the tradeoff between using crop residues as animal feed or for soil management. This study would have an impact on the water balance of the whole basin even though the adoption would be at farm level.
- Assess which livestock species and breeds could improve water productivity (in economical as well as environmental aspects) for the different environmental niches especially in water stressed areas.
- Optimizing irrigation schemes by growing suitable forages (near main and secondary canals for minimizing evaporation) for animals without reducing the water available for the main crop production.
- Assess the socio-economic impacts of changes in the traditional systems where the role of livestock is intensified or diminished. Livestock water productivity could be one of the major assessing criteria.

Feed Sourcing Strategies: Food crops are grown for sustenance and income. Efficient use of the crop residues and by-products of those food crops for value-added animal production increases food production and incomes without significantly increasing use of transpired water. Well-managed grazing makes efficient and productive use of rainfall on land unsuitable for crops. Overgrazing compacts soils reduces plant diversity and increases surface water run-off. Optimal grazing conserves soils and encourages infiltration of rainwater that helps prevent down-slope, flooding and sedimentation of reservoirs. In contrast to grazing livestock on the stubble that remains after harvesting, cut-and-carry feeding methods combined with conservation tillage reduces water run-off and soil erosion associated with annual crop production. Application of manure also increases the water-holding capacity of crop soils.

- Integration within cropping systems, utilization of residues for feed, manure inputs.
- Utilization of food-feed crops instead of sole purpose forage crops.
- Use irrigation infrastructure for feed production particularly where animal power is needed for cultivation.
- Import feed if water (virtual water) is scarce or highly valued.
- Feeding of balanced diet.
- Improvement of poor quality roughages.

Rain fed livestock production in rangeland areas: Encourage collective action to reduced stocking rates.

- Keep animal numbers at or below carrying capacity.
- Increase water productivity per animal.
- Establish policy and encourage livelihoods linked to reallocation of transpired water to restore agro- ecosystem diversity, stability, and resilience.
- Encourage socially acceptable and guaranteed alternate wealth savings.
- Encourage collective strategies for reducing drought risk.
- Balance mobility of use of grazing resources with supply of drinking water.

Water Management Strategies

a. Drinking water

- Provide sufficient quality water for optimal feed conversion efficiency and maximum growth and milk production
- Reduce walking distance to water source or deliver water to animals
- Combine with zero-grazing strategies
- Integrate livestock watering infrastructure into water development for domestic and irrigation needs
- Develop drinking water sources to allow utilization of previously unused rangelands
- Constrain herd sizes and animal movements so that land and vegetation adjacent to watering point is not unduly degraded
- Separate livestock watering from water source used by people for domestic consumption
- Keep livestock out of water reservoirs to prevent erosion, sedimentation, and spread of water-borne diseases involving livestock.

b. Soil and ground water conservation in mixed crop livestock systems

- Encourage conservation tillage including restriction of animal feeding on residues to maintain adequate ground cover and to reduce the number of oxen needed
- Encourage zero grazing
- Promote terracing to increase food-feed crop production
- All water conservation measures for crop production
- Recycling of waste water.

c. Animal Productivity Enhancing Strategies

- Breed
- Health care
- Reproductive care and management
- Scientific management practices

Marketing chain and value addition of livestock products

Understanding the diverse roles of livestock in culturally and environmentally varied agricultural production systems in India, it appears that there are many opportunities to greatly improve livestock water productivity and the contribution of livestock keeping to overall water productivity. In using water for agricultural production, one challenge is to objectively assess the potential of contribution of livestock and livestock products as one of many agricultural products that can be produced per drop of water depleted. There is also an ample scope for improvement in water productivity in crop-livestock system in semi-arid region of India where productivity of both crop and livestock enterprises is low, partly because of negligence on crop-livestock and water nexus and their impact on livelihood.

Water Productivity of Aquaculture

The Eastern Region of India is bestowed with enormous aquatic resources. This region is having around 0.65 Mha ponds and tanks, 0.56 Mha of reservoirs and 0.48 Mha of flood plain lakes. With the vast aquatic resources there has been a continuous increase in fish production during last decade and contributes around 33.98% of the total fish production of India and 48.85% the total inland fish production of India during 2012-13.

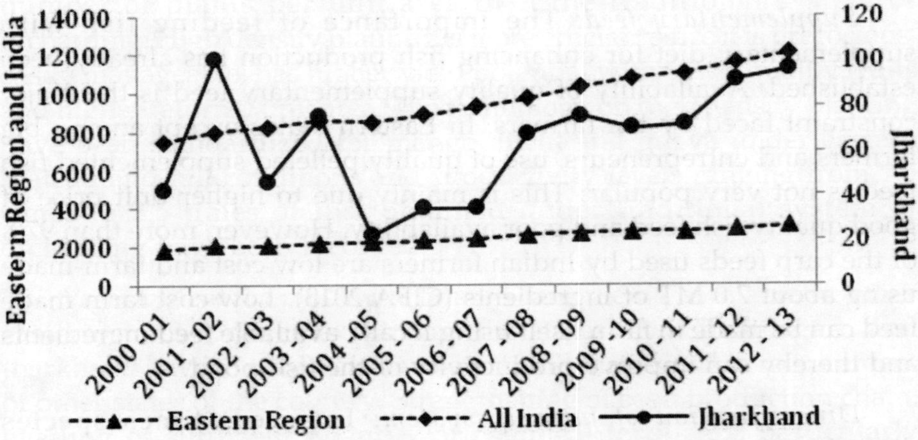

Fig. 8.2 Trends in fish production of Jharkhand and Eastern Region (000' ton)

In spite of vast aquatic resources, productivity from this region is less compared to many other states of India like Punjab. Moreover, in-spite of having tremendous potentiality, productivity of the open water bodies are still very meager. With the growing demand of fish, it is very important that open water bodies like flood plain lakes, beels, reservoirs etc also get due importance and thereby productivity and overall production from these water bodies can be increased through optimize management practices. Some of the strategies for increasing production and productivity of the water bodies are listed below:

Improving productivity of Freshwater fish culture: As per the FFDA report average productivity of India is around 2.9 t/ha where as it is 2.3t/ha in Jharkhand and 2.4 t/h in Bihar and as high as 4.4 t/ha in West Bengal. Improvement in productivity of the existing pond based farming system through scientific method of farming, diversification of cultured practices; selective stocking of desirable species and size at stocking, proper water quality management, intercropping with minor carps etc. can help in improving productivity of the existing system.

Quality fish seed for farming community: Quality fish seed for stocking is the major problem in the Eastern Region. The region houses hundreds of hatcheries (Jharkhand is having 31 hatcheries) but quality of the fish seed is not up to the mark. Healthy, genetically superior, disease free and appropriate size and age of the seed are some of the quality parameters that need special attention.

Supplementary feed: The importance of feeding fish with supplementary diet for enhancing fish production has already been established. Availability of quality supplementary feed is the major constraint faced by the farmers. In Eastern States except among big farmers and entrepreneurs, use of quality pelleted supplemented fish feed is not very popular. This is mainly due to higher unit price of good quality fish feed and poor availability. However, more than 97% of the carp feeds used by Indian farmers are low cost and farm-made using about 7.0 MT of ingredients (CIFA 2013). Low cost farm made feed can be made in farm itself using locally available feed ingredients and thereby can improve productivity of the fish pond.

Diversification of farming system: In recent times species diversification has proved to be more rewarding among farmers in terms of higher total yield and income. Intercropping with minor carps (like bata, gonius, reba), Java puthi, air breathing fishes, prawn as well as other smaller self-recruited species etc. in polyculture can be more remunerative. Production of these water bodies can be further enhanced by adopting non-conventional method of culture practices like cage and pen culture practices. In this regards Jharkhand has made significant contributions especially through cage culture practices.

Fish health management: With the intensification of the culture practices like high stocking density, unbalanced diet, poor seed quality, poor water quality, high organic load, occurrence of large numbers of disease causing agents, pesticides in water etc. fishes are always under stress and becoming vulnerable to disease. Hence, proper health management practice has to be adopted so that stress will be minimized and growth will be maximized. Quality fish feed in required quantities, quality fish seed, optimum water quality parameters, proper prophylactic measures can be wholestic approach to check the horizontal as well as vertical transmission of disease.

Integrated fish farming: Integrated fish farming system holds great potential in Eastern Region for not only augmenting total fish production but will also generate additional income and employment. The main objective of this system of farming is to recycle wastes or by-products (faeces, urine and spoiled feeds) generated from one unit can be used as inputs to fish culture in the form of fertilizer, feed etc. so as to achieve higher productivity and to reduce the cost of production. This region is having large livestock and avian population. Hence, efficient utilization of these dung and agriculture waste for production purpose through an integrated approach will help in recycling the waste generated, simultaneously will solve pollution related problems.

Marketing system: Marketing of fish and fishery products is still not an organized sector and running without any proper channel and system. Fish being a perishable product proper infrastructure like hygienic market; transportation & distribution through cold chain system; development of market research group for proper pricing are some of the challenges to achieve over all development in fisheries.

Development of location specific and farmer's friendly technologies suitable for the region will boost the fish production of the region. However, there are several issues hampering the development and improving productivity from open water bodies, marshy lands, derelict water bodies, flood plain water bodies and reservoirs like inadequate knowledge about the productivity status of the water bodies, non-availability of quality fish seed, high macrophytes infestation, poor auto stocking due to heavy siltation in the river connection, non-selective method of fish harvest, poor technological support, inaccessibility and transportation facilities etc.

Improving Water Productivity of Poultry

Integrated farming system aims at delivering more sustainable agriculture with an emphasis on holistic management approach looking at a whole farm as cross-linked unit. Integration of various agricultural enterprises like crop, animal husbandry, fishery, poultry, forestry etc. has great potentialities in the agriculture economy. These enterprises not only supplement the income of the farmers but also help in increasing the farm labour employment. The soil type, rainfall distribution, length of growing season, availability of resources in the system have to be considered while selecting any new component for optimum water productivity.

Integration of poultry in land use based farming system is right answer for sustainable and eco-friendly development of agriculture for optimum water productivity as well as need of the region point of view. The rural poultry farming can also play an important role in alleviating protein mal-nutrition, poverty and to generate employment for tribal women and unemployed youth too. The commercial poultry production has reached the stage of stagnation, but rural poultry farming is still having ample scope to fight against protein mal-nutrition due to an acute shortage of eggs and poultry meat (GOI, 2014). The annual per capita egg availability in our country is 58 whereas; in the state of Jharkhand is only 13 numbers (Table 8.3).

Prospect of Poultry Farming

Total poultry population in our country is about 729 million and out of 105 million under backyard system with concentration of around 84 percent of *Desi* type. Almost 65% of eggs and 70% of poultry meat produced in our country are consumed by only 1/3rd of the population residing in urban areas and rest of people who lives in rural areas are deprived (Churchil and Narayanankutty, 2003). On the other hand, rural poultry farming is patronized generally by poor and weaker sections of societies. Even today more than 40% of poultry population in the country consists of different types of indigenous chicken that thrives only on scavenging. Improved breed of rural poultry like

Vanaraja, Gramapriya, Gramalakshmi, Giriraja, Girirani, Krishna, and Ganga etc. are available in India. These breeds of poultry are having capacity to produce 150-180 eggs on minimum inputs in free-range system of production against 70-100 eggs by *Desi* non-descript poultry. Therefore, emphasis must be given for popularization of rural poultry in our country to improve socio-economic status of poor people. Thus, rural poultry production under semi-intensive or intensive system as well as broiler chicken farming under integrated farming system is viable component to improve water productivity through multiple uses. For better management and efficient uses of local resources are explained below for more return.

Table 8.3 Poultry Scenario of India and Jharkhand

Particulars	India	Jharkhand
Total poultry population	729.21	13.56
Rank in the world	5th	–
Backyard poultry population	104.72	5.02
Population of *Desi* bird under backyard system (%)	84.25	91.23
Annual egg production	69.73	0.42
Rank in the world	3rd	–
Annual per capita egg availability (Nos.)	58	13

General Management Practices

Improved rural breeds or broiler strain selected for integration in the farming system should be hardy and thrive well on locally available feed resources. Birds may be maintained under semi-intensive or intensive system of rearing. The locally available materials like bamboo, wooden planks, and thatch grasses may be used for the construction of house at elevated places. The details of space requirement are given in Table 8.4.

Table 8.4 Space Requirement of Poultry

Stage	Floor Space	Feeder Space	Waterer Space
Brooding (0-8 wks.)	450	2 - 7	0.5 - 1.5
Growing (8-18 wks.)	900 - 1100	7 - 10	1.5 – 2.5
Laying (<18 wks.)	1800 - 2200	12 - 15	2.5
Broiler Finisher	750 - 850	7 - 10	1.5 – 2.5

Poultry is warm blooded animal and they feel most comfortable at temperature of about 23-27°C. However, at day-old age, the brooder house temp. Is required 35°C and it shall be decreased @3°C per week to the level of 23-27°C. The saw dust, paddy husk or groundnut hulls can be used as bedding (7-10cm thick) materials. Mixing of lime @10kg/100ft^2 is helpful in reducing moisture level and ammonia gas from litter. Proper ventilation is also required in house.

Birds are allowed from early morning to evening for scavenging in side fenced area under semi-intensive system of rearing with free access of clean water. Household waste, crop by-products, broken grain etc. are alsooffered to the birds depending upon their nutrient requirement. However, birds under brooder or intensive system areprovided compound mash/pellet feed 3-4 times as per their requirement (ICAR, 1985). The major nutrient requirement of poultry for different stages with some feed formula for preparation is given in Tables.

Table 8.5 Major nutrient requirement of poultry

Nutrients	Broiler		Layer		
	Starter (0-4 wks.)	Finisher (5-8 wks.)	Starter (0–8 wks.)	Grower (9-20 wks.)	Laying Period
Protein (%)	22	19	20	15	18
ME (Kcal/kg)	2900	2900	2700	2400	2600
Fiber (%)	6	6	6	6	6
Calcium	1	1	1	1	3.5
Available Phosphorus	0.5	0.5	0.5	0.5	0.5
Lysine	1.22	1.06	1.22	0.85	0.50
Methionine+Cysteine	0.83	0.72	0.83	0.58	0.54

Table 8.6 Poultry feed formula

Feeds (kg) with CP & ME Values	Starter	Finishe	Layer
Maize: CP 8.8%, ME 3330Kcal/kg	50	60	50
Rice Polish: CP 12%, ME 2850Kcal/kg	10	10	20
GNC: CP 40%, ME 2600Kcal/kg	22	15	10
Soya Meal: CP 42%, ME 2700Kcal/kg	8	5	8
Fish Meal: CP 45%, ME 1850Kcal/kg	8	8	9
Additives (Vit./Minerals): CP/ME: Nil	2	2	3

Generally one broiler chicken can consume total 4.5-5.5kg compound feed in their life period of 6-7 weeks. Compound concentrate feed can also be prepared at house by mixing locally available non-conventional feed resources like damaged cereal grain, soybean, rice bean, Job's tears, buckwheat, jack bean etc. up to some extent as given below to make economic ration since it represents 65-70 percent of the total cost of production (Gupta and Bujarbaruah, 2005). Farmers always must try to control mortality and in any case it should not increase over 10% for economical production. For this, all general health practices should be followed like vaccination, treatment of water and litter. The following preventive measures to be strictly followed.

Table 8.7 Vaccination Schedule

0-3 days	Ranikhet / Fowl Pox
5-6 weeks	Second dose of Ranikhet / Fowl Pox
Preventive Measures	
0-3 weeks	B-complex
4-5 weeks	Coccidiostate
5-7 weeks & During Laying	Calcium tonic fortified with B-12

Economic Feasibility of Broiler Chicken Production

The economics of broiler chicken production are calculated on assuming present market prices are presented in Table 21.5, which show good return and that ultimately enhance water productivity of the system.

Table 8.8 Economics of Broiler Chicken Production

Recurring Expenditure: 500 Birds (Rs.)	
i. Stock (day old Chicks @ Rs. 24 each)	12,000.00
ii. Feed @ 4.2kg/chick for 60 days @ Rs. 22/kg	46,200.00
iii. Veterinary medicine, litter etc. @ Rs. 10/chick	5,000.00
iv. One labour @ Rs. 200/day for 60 days	12,000.00
Total *(Expenditure per bird Rs. 150.40)*	75,200.00
Income: (Rs.)	
i. Sale of live broiler after 60 days weighing average weight 2kg and sold @ Rs. 90/kg live wt. and considering 5% mortality	85,500.00
ii. Sale of poultry litter (App. 2 ton) @ Rs. 2000/ton	4,000.00

iii. Sale of empty feed bags (app. 40 Nos.) @ Rs. 5/each	200.00
Total	89,700.00
Profit	
(Profit/bird Rs. 29.00)	14,500.00

Integration of suitable rural poultry breeds in farming system or in watersheds is certainly feasible to fight against protein mal-nutrition and at the same time to improve water productivity of the system as well as to improve the economic and health status of poor people in particular.

Strategies to Improve Water Producitivity of Duckery

Duck is the second important avian species of economic importance next to chicken. Commercial duck farms are not found in our country. Keeping the ducks in confinement with commercial feed may not be economically sustainable always, especially for the small and marginal farmers. But rearing in extensive management in village ponds or with semi-intensive management is proved to be a profitable and viable means of livelihood for rural farmers.

Advantages of duck rearing

- Ducks utilize the feeding source from both land and water. They have remarkable ability to survive on feeding materials that are not retrievable by chicken, turkey or other domestic animals.
- Ducks lay more eggs, comparatively bigger in size and can lay in subsequent years of production. Productive life span of ducks is more than that of chicken.
- Duckery does not require high investment on infrastructure development like housing etc.
- Ducks are hardy as they are more resistant to many diseases as compared to chickens. Ranikhet disease and Avian influenza have not been reported in this species.
- They get their feed from natural source i.e., pond, marshy lands etc. where other domestic animals cannot be reared.
- These birds have better adaptability to the environment.
- Mortality due to predator is less in ducks as they prefer to remain more time in water.

- Duck rearing can be well undertaken as a group activity particularly by the women farmers.
- Duck meat can fetch good market on account of its quality.
- Ducks can be integrated with other agricultural practices like aquaculture, rice cultivation or both.

Khaki Campbell *White Pekin*

Native cross-bred duck *Duck integration in paddy*

Fig. 8.2 Different breed of duck and integration in paddy fields

Starting a duck unit: A duck unit can be started by the farmer during any time of the year. However, the minimum requirements to initiate the practice need to be available.

Construction of duck house: A low cost house needs to be constructed with locally available materials like bamboo, mud, thatching straw etc. Floor space required is 2-0 to 2.5 sqft/bird. Location of the duck house preferred to be nearer to pond but secured from predators. The house should be well ventilated. For adult ducks sand may be considered as cheap and suitable litter material. But during brooding, rice husk may be used which need to be replaced frequently. In some occasions wooden house fitted with wire net in all sides and thatched with tin or asbestos sheets are constructed for better hygienic condition with more durability.

Day old ducklings: At present three breeds of duck are available for propagation in farmer's field. 1) Khaki Campbell: this layer breed of ducks is well known for their tremendous potential of egg laying (280-320 eggs/year). However, they are poor scavengers and need supplementary diet for optimum production. 2) Native cross bred ducks: They have multi colored plumage and are considered a good layer breed (220-260 eggs/year). They are very good foragers and scavengers with better adaptability than other breeds. 3) White Pekin: White in color, heavy weight, fast growth rate, poor laying ability and is a popular meat breed (>2.5 kg in 6 wk age; 120-140 eggs/year). Due to the fast growth these birds need sufficient food balanced with protein and energy. They produce one of the best quality duck meat in world.

Brooding of day old ducklings: Ducklings hatched in the same day are transported to different parts and brooded in warm environment for nearly 7-10 days. Duck mash (*ad libidum*) soaked in water is provided 2-3 times a day as starting ration. Sufficient clean drinking water is offered round the clock. Electric bulb (100 -200 watt) fitted with reflector is to be fixed at a desired height so that the ducklings are exposed to warm environment. Care should be taken to keep the floor / litter dry during brooding period.

Management of growing and adult ducks: Ten days brooding is sufficient for ducklings of any breed. After this, they are allowed to scavenge outside. Commercial food is not needed after brooding period. Sufficient water in plastic troughs may be offered so that the ducklings can dip their heads to avoid blindness in the flock. Kitchen waste, left over food of the family, poor quality rice (boiled) need to be offered to meet nutritional requirement of these growing birds. After one month of age, ducklings are to be allowed to enter the pond and care must be taken to prevent casualty or mortality due to predators during this critical stage of life. It is expected that the laying variety of ducks will attain more than 1kg body weight by 10th week of age and the meat variety may cross 2 kg body weight by 8th week age.

Identification of male and female ducks in a flock is important as adult drakes (male duck) consume more food material from the surrounding. Due to shortage of food there may be delay in sexual maturity of female ducks resulting in delayed egg production. So drakes in the flock need to be identified and disposed by 3-4 months of age for meat purpose that fetches a good income for the farmer. Female ducks quack more as compared to the males. Feathers on the tail are prominently curled in male ducks. Sometimes a band like colouration in the neck is found in the male ducks especially in Khaki Campbell and Cross-bred Native ducks. Drakes are comparatively heavier than ducks.

Under free range, ducks (Khaki Campbell/ Crossed Native) usually lay eggs around 20 weeks of age. They usually lay eggs during the late night to early morning, which makes it easy to collect all the eggs from the duck house. Sand as litter material and wide mouthed earthen pots placed inside the house facilitates clean egg production. Calcium supplementation during laying period is necessary for continued egg production in the flock. Simultaneously, care should be taken to provide sufficient feeding source for the female laying ducks for higher productivity. The average size of duck egg is above 60g which fetch higher market price in comparison to chicken eggs.

Health care management

Ducks are hardy as compared to poultry. Change in climate for a long period and non-availability of water during summer may lead to reduced egg production and poor health of the bird. Duck plague is the most serious viral disease in ducks observed with symptoms of profuse diarrhea, high fever and hemorrhagic spots in digestive tract. Mortality is quite high. Vaccination against duck plague during 3-4 months of age once in a year may eliminate the possibility of its outbreak. Duck hepatitis is another viral disease which affects the liver of young birds of 4-6 weeks of age and may cause high mortality. Vaccination at 2-3 months of age is needed to protect the birds. Besides, aflatoxicosis (fungal origin) is a common problem observed in organized duck farms offering commercial feed which can be corrected by providing freshly prepared feed regularly.

Integrated duck production

Duck farming can be integrated with aquaculture for better profitability. Droppings of ducks are rich in nitrogen and phosphorus which promotes plankton production in the pond needed for fish consumption. Due to continuous swimming of ducks the temperature of pond water remains uniform. On the other hand ducks collect their food from the pond and use the water for its physiological needs. Thus the farmer gets benefited from both the components in terms of productivity and employment generation. Similarly duck production can be integrated with fish and rice cultivation for further economic gain.

Economics of duck farming: Duck rearing is the alternative to poultry farming. Farmers can use their water bodies for duck as well as fish production. Even in the absence of ponds, a polythene pond can be prepared and 20-30 ducks can be raised with home waste feed materials. Egg and meat are always in demand which provides regular income to a farmer. It was calculated that a farmer family could generate rupees 12,000-13,000/- per year from a flock of 25-30 numbers of ducks which provides a great aid to his livelihood.

Thus, extensive duck rearing for meat and egg production by small, marginal and progressive farmers need to be encouraged by various organizations through training, demonstrations and capacity building etc. Involvement of self-help-groups in this regard is essential. Organizations should look forward for availability of critical input at the farmer's door step at right time with reasonable price. Above all, the role of KVKs for promoting duck rearing is need of the hour for realising "Egg surplus India" in the future.

Integrating Pig in Agricultural System

Piggery is a highly profitable enterprise because of among the domestic animals, pigs are the most prolific. Feed conversion ratio is 3.0 to 3.5 kg feed per kg live weight gain. In India, pig farming and pork production are in the hands of traditional pig farmers belonging to lower income group people. The food habits of Indians are changing and there is a great demand for pig meat. Integrated pig farming with fish is a demanding venture of enhanced income in the family of lower strata people.

Pig can fit into diverse systems of management. It can be reared on kitchen wastes, industrial or agricultural by-products and slaughter house offal. Pigs are voracious eater and can convert agro-waste in to edible and highly nutritious meat. One of the best feed converter and prolific amongst the all domesticated livestock. With short generation interval a sow can produce 16 piglets in a year. Dressing percentage is very high (65 to 70%) in comparison to sheep-goat (45 to 55%). Piggery can generate employment opportunity for rural population.

Several improved breeds of pig have been developed. Yorkshire, Barkshire, Hampshire, Duroc, Hereford, Landrace, Chester white, Tamworth, Tamworth x Deshi are some of the breeds suitable for possible inclusion in to integrated farmings systems.

Management of pig in IFS

Breeding management: Pigs can be bred at the age of 8 to 9 month or 80 to 90 kg body weight and the boar-sow ratio should be 1:10. Generally, a sow can be bred 3 times in 2 year. Oestrus cycle in pig is 21 days. Gestation period in pig is 114 days. Generally the litter size is 8 to 10 numbers.

Feeding of pig: Feed may be provided two times in a day. Morning and evening feeding is preferred. A 70 to 80 kg body weight mature pig consume 2.5 to 3.0 kg feed per day. To reduce feed cost swill feeding or kitchen waste or agri- waste may be provided to pig.

Health management: Pigs are susceptible to heat stroke. Provision of wallowing or water sprinkling is very important during summer months. Vaccinations for Swine fever and Food and Mouth disease must be done. Regular deworming is also advised.

Potentiality of pig manure: The average dung productions in pigs are 0.25 tonnes per pig per year. Pig manure is also a good source of nitrogen and phosphorus. In pig excreta N, P, K contents are 0.60%, 0.50%, 0.20%, respectively. Pigs under 50 kg body weights on average produce 2.5 kg dung per day and 50 to 90 kg pig produce 5 kg per day and it contains 20 to 25% dry matter.

Integrated pig with fish farming

In integrated farming system fish cultured for direct consumption of pig manure or from spilled over feed. Pig components help in the generation of fish food organisms (planktons) which form the primary feed for fishes. The integrated fish farming may be extensive, semi intensive or intensive sub system. Excess manure may lead to poor water quality, leading to depletion of dissolved oxygen level and fish mortality.

In pig-fish integrated farming system the growth rate (g/fish/ day) of Catla, Grass Carp, Common Carp and Mrigal is 1.29, 0.87, 0.50 and 0.80, respectively. The economic return in integrated farming system with livestock like pig-fish farming system in first year in terms of input- output ratio is 1:1.95 and 1: 2.11 in pig- duck-fish system. In second year the ratio is 1:2.23 and 1:2.42, respectively.

❑❑❑

9

Crop Diversification Options

Diversification is one of the components of agricultural growth in addition to area expansion and technology development. Crop diversification assumed high importance in the recent past due to two reasons; fatigue in other components of growth in Indian agriculture and new opportunities for diversification towards high value crops due to the demand and supply side factors. While, taking diversification decisions, farmers attach more importance to expected total net income from the crop and hence consider both the price and yield. Relative incomes from the crops explain the crop substitution decisions of farmers i.e., farmers calculate the aggregate gain from the crop than considering only the price of the crop. The capacity to generate higher productivity and availability of better marketing prospects together explains farmers' decision. In terms of risk, farmers are concerned not only about crop-specific risk but also their aggregate production and consumption risk. So improving farmers' orientation towards high value crops require intervention in food grain markets by improving the technology.

Better technology in food crop can lead to increase in the productivity of food crops which would help in reducing consumption constraints for higher allocation of land to high value crops like horticultural crops. Recognizing that the Indian economy is agriculture-oriented, some perceptible changes during last three-and-a-half decades have impacted the agricultural growth and the overall development patterns of the economy. The size of operational holdings in India has declined from 2.28 ha in 1971 to 1.57 ha in 1991 to 1.41 ha in 1995-96 and some estimates say it has further declined to 1.22 ha in recent years, which has an impact on crop productivity. Future increase in agricultural growth has to be essentially achieved through increase in yields. Cereal yields have gone down and at the same time consumer preferences have shifted away from cereals and moved towards high-value agricultural produce. Higher incomes and urbanization in India, changing lifestyles, international market integration and trade liberalization are expected to increase the demand for high value agricultural produce like horticultural products rapidly.

Diversification to Horticultural Crops

Horticultural crops have played major role in increasing income of farmers through crop diversification during the last decade. A part of the remarkable growth rate of horticulture sector during the past 10 years is attributed to government initiatives taken during the period. Horticulture sector has received focused attention in our planning process from 7th five year plan period onwards; as a result, there has not only been sustained increase in production of horticulture crops but hi-tech horticulture has also been recognized as a commercial proposition. It is a fact that horticulture sector has provided opportunity of crop diversification, resulting into increased income from the land and also the nutrition security. The benefit of area expansion in horticulture in clusters supported by post-harvest management infrastructure has percolated down to even small and marginal farmers a number of whom contribute to exports of horticulture produce too.

Since horticultural crops are labor-intensive and generate high and quick returns, these provide farmers a chance to utilize their surplus labor and augment their incomes. Horticultural crops have been acclaimed as a special case for diversification on several other counts as well. The increased trend towards specialization in cropping pattern in few states coupled with stagnancy in their yields and increasing cost of production is bringing down the real income of the farmers. This highlights the need to look for alternative cropping pattern with a view to remedy the situation. Due to higher returns and productivity of horticultural crops, this group has emerged as an important area for diversification and as an alternative cropping pattern. Available evidence shows that one hectare under horticultural crops can generate an annual income up to Rs. 20,000, compared to hardly Rs. 10000 and Rs. 4000 by rice and ragi respectively. Data given in Table 9.1 indicates higher values of benefit cost ratio in case of horticultural crops over that of foodgrain crops. Cultivation of fruits can generate employment to the tune of 860 man-days as against 143 man-days for cereals. Diversification towards horticultural crops is expounded as the strength of Indian agriculture, due to their potential to produce a wide variety of horticultural crops under varied agro-climatic conditions.

Horticulture based production system for crop diversification

An appropriate crop diversification option, particularly under the eastern plateau and hill conditions should, not only be economically profitable (based on common methods of estimation in rupee term), but also be a water-use-efficient system and have higher value of ecosystem services. Contrary to the common belief of high water requirement of horticultural crops like vegetables, the data in Table 9.2 indicates higher values of water use efficiency of horticultural crops than the grain crops.

Table 9.1 Benefit cost ratio of different crops

Crop	State	Benefit cost ratio
Wheat	Bihar	0.91
	Gujarat	1.30
	Haryana	1.16
	Madhya Pradesh	0.92
	Punjab	1.23
Paddy	Andhra Pradesh	0.94
	Bihar	0.88
	Haryana	1.18
	Madhya Pradesh	0.74
	Punjab	1.42
Bajra	Gujarat	0.78
	Haryana	0.46
	Maharashtra	0.50
	Rajasthan	0.56
	Punjab	1.42
Maize	Andhra Pradesh	0.88
	Bihar	0.75
	Karnataka	0.83
	Rajasthan	0.45
	Uttar Pradesh	0.63
Mango	Uttar Pradesh	4.60
Litchi	Bihar	2.61
		1.69
Guava	North India	1.95
Brinjal	Andhra Pradesh	1.98
Bhindi	Karnataka	1.71
Tomato	Karnataka	2.01

*Definition of cost include cost value of all inputs, depreciation, rent and implicit value of land and family labour. *Source:* Mittal (2007)

Horticulture based production systems with perennial plant as integral components are the most effective options for crop diversification through horticultural intervention in the ecologically fragile situation in eastern plateau and hill region. Land cultivated in this way can maintain an ecological balance and provide for efficient use of all natural resources. Land use systems based on mixed tree crops have clear advantages over annual cropping systems for the maintenance of soil fertility in the humid tropics.

Table 9. 2 Water use efficiency of different crops

Crops	Water use efficiency (kg/mm water)	Crops	Water use efficiency (kg/mm water)
Bottle gourd	45.25	Sorghum	9.00
Brinjal	31.11	Bajra	8.00
Onion	15.45	Maize	8.00
Tomato	12.41	Wheat	12.50
Papaya	5.69	Gram	3.66
Banana	32.67	Ground nut	4.33
Lemon	44.76	Rice	3.70
Mosambi	60.24		

Source: nrlp .iwmi. org

The advantages include permanent soil protection; a more favorite environment for soil biological processes, which affect litter decomposition and soil structural improvement, and more efficient nutrient cycling. Therefore, the integration of deep-rooting trees into the production system increases the nutrient availability in the topsoil, if the quantity of nutrients taken up from below the crop-rooting zone is greater than the quantity stored in the tree biomass and undecomposed tree litter. The fruit based multitier cropping system is one of such technologies which have been successfully standardized and translated into farmers' fields for increasing the profitability in the uplands of the region. The Fruit based multitier cropping system is can be defined as a planting system comprising combinations of plants with various morpho-phenological features to maximize the natural resource use efficiency and enhanced total factor productivity.

The multi-tier cropping system comprise of a combination of perennial and annual plant species as different components in the same piece of land arranged in a geometry that facilitates maximum utilization of space in four dimensions (length, width, height and depth) leading to maximum economic productivity of the system. Farmers in states like Kerala, West Bengal have been following this practice since long time with coconut or arecanut as main crop. Research conducted on different fruit based multitier cropping system have indicated a productivity level of 8 to 10 tonnes per ha against 1-2 tonnes per ha under traditional system of monocropping in the plateau regions of eastern India. In studies conducted at Himachal Pradesh, CO_2 mitigation level of Hortipastoral and hortisilvipastoral system have been estimated to be 33.87 and 40.40 t/ha, respectively which were much higher than that was recorded under natural grass land (7.34 t/ha) indicating the ecosystem services of the tree based tree based system (Khaki and Wani, 2011).

A number of other horticultural technologies have been standardized at research institutes and SAUs working in the area for increasing the agrarian profitability through crop diversification under the eastern plateau and hill region. Some of the promising technologies are given below;

High density planting (HDP): High density planting is another approach for maximization of profitability through increasing the land use efficiency. It is the practice of growing plants at a closer spacing than the traditional system of orcharding to accommodate more number of plants per unit area than the traditional system of orcharding. High density planting is more efficient since it is precocious, easily manageable, has higher yield potential with better quality fruits and higher returns/unit area. The technology on high density planting have been standardized for mango, litchi and guava under eastern plateau and hill conditions and productivity levels ranging from 1.5 times to 2.5 times of that obtained under traditional system of orcharding have been recorded.

Vegetable cultivation: With limited land being available in other eastern states for area expansion, the states like Chhattisgarh and Jharkhand have to play a major role to fulfill the vegetable requirement of other states of the country. Strengthening of seed production chain, availing of sufficient quantity of chemical fertilizers, particularly potassic fertilizer and micronutrients, creation of strong infrastructure for post harvest handling of vegetables including long distance transport will go a long way in fostering a vegetable revolution in the region.

Vegetable cultivation with adoption of resource conservation practices like drip irrigation and mulching have shown significant increase in the productivity under the eastern plateau and hill conditions. The ICAR Research Complex for Eastern Region has released 37 varieties/ hybrids of different vegetables viz. tomato, brinjal, cucumber, ridge gourd, sponge gourd, pumpkin, garden pea, cowpea, French bean, snow pea, vegetable, soyabean, dolichos bean for cultivation in the region.

Round the year production of mushroom: The region being predominantly a paddy growing region, offers ample scope for mushroom cultivation due to the availability of crop residue. Among the different species of oyster mushroom, *Pleurotus florida* can be grown during August-November and January-May, *P. sajorcaju*and *P.flabellatus* can be grown during August to November and January to June, *P. ostreatus* can be grown during Nov-March and Blue oyster mushroom (*Hypsizyus ulmarius*) can be grown during November to January. The milky mushroom (*Calocybe indica*) can be grown during April-August in the same culture room with provision of blue coloured light at the time of fruit initiation.

Under eastern plateau and hill condition, horticulture based cropping systems can effectively address the issues on water use efficiency, minimization of soil erosion, enrich biodiversity, provide ecosystem services, higher income, support employment generation, promote farm level innovation and gender equity. Horticulture can be promoted in a holistic way as a means of agro-diversification for the second Green Revolution, providing the much needed impetus to the growth of agricultural sector, through increase in trade, income and employment.

Diversification to Medicinal Plants

Medicinal Plants are those that are commonly used in treating and preventing specific aliments and diseases, and considered to play a beneficial role in health care. The medicinal plants are the living resources; if used in a sustainable manner, will serve the mankind for a long period. The forest area has been the traditional source for plants and herbs, but due to destructive & unsustainable methods of harvesting, the resource base have been depleting very alarmingly. In contrast, the demand for medicinal plants is increasing in both developing and developed countries due to growing recognition to natural products, being non-narcotic, having no side effect and some time may be the only source of healthcare. In the last three decades, several institutes initiated R&D work on medicinal plants and brought out useful information on crop improvement, post-harvest, development of improved varieties and analytic procedures of several medicinal plants. Now the area under medicinal plants cultivation is expanding at faster rate and farmers are asking for quality seed and planting material along with the agro technology.

India with its rich biodiversity and varied agro-ecological setting is an ideal habitat for a large number of medicinal & aromatic plants. Plants are intimately associated with the life of men and other animals fulfilling all their daily needs. But unscientific exploitation of plants wealth especially from forest areas causes stress in biodiversity and many of the plant species are on the verge of extinction. This call for integration of efforts for conservation and commercial exploitation of such plant species. Technological development in genetic engineering and biotechnology and new screening producers to detect active principles, have stimulated widespread interest for medicinal plants. Therefore, concerted, sincere and honest efforts for systematic survey, collection, inventory of diversity of MAP and search for new plant/ species and their associated knowledge are required to propagate this wealth.

The use of medicinal plants for treatment dates back to over 5,000 years and has been codified in the form of Ayurveds, Unani and Sidha medicines. The World Health Organization (WHO) has compiled a list of 20,000 medicinal plants, used in different parts of the world. India has ancient history of plants use in the indigenous system. Due to extreme and diverse topographic and climatic conditions, India is rich in medicinal and aromatic plants. About 2500 plants having medicinal value have been officially recognized and are being used by the local tribes and villagers since prehistoric times. Jharkhand is a State full of forest wealth and is one of the hotspots of plant biodiversity. Medicinal and aromatic plants are treasures of the Jharkhand forests which make significant contribution in boosting the economy of the people in forest region. Most of the plants/plant parts in the international trade are wild collection from the forest.

Indian Scenario of MAP wealth

In Indian medicines around 25000 effective plant based formulations are available and being manufactured by Ayurvedic,

Unani, Siddha, Homopathy, Homeopathy, and cosmetic industries. Over 1.5 million practitioners of the Indian system of Medicines use medicinal plants in preventive, promotional and curative applications. It is estimated that there are over 7800 medicinal drug-manufacturing units in India which consume about 2000 tonnes or more of herbs annually. Medicinal plants as a group herbs, shrubs, climbers, lianas trees and epiphytes comprise approximately 8000 species and account for about 50% of all the higher flowering plant species used for the production of medicine by the Indian herbal industry, less than 20 species are currently under cultivation in different parts of the country. In the country around 70% of the medicinal plants are spread across the tropical forests of western Ghat, Vindhyas, Chhotanagpur & Santhal Pargana plateaus, Aravalli, Terai region, Himalayas and North East. Indian climatic conditions having wide ecological amplitude is suitable for cultivation of medicinal plants which provided a great share in the Indian system of medicine, which is slowly developing into a form of alternate medicines globally.

Problems and Prospects

The rural population not only depends on medicinal resources for health needs but also for their livelihood generation. But due to lack of knowledge about phenological pattern of plant, methods of propagation, and having no direct marketing system, the villagers receive very meager and middle men earn a lot of money by selling them with big firms. Hence, strengthening of the segment through wider marketing linkages, better quality produce and increase in productivity will improve the rural economy as well. The opportunity in this sector, in terms of medicinal plant cultivation and trade is enormous, due to huge demand at country level and abroad. The cultivators are this sector is very few and they need further refinement in their agro-technique. The cultivators sell their product mostly through individual initiative and not through organized channel, thus they are getting lower price. The advantages of high price offered by institutional buyers are mainly received by middlemen. Hence market-mapping activity is essential for healthy growth in medicinal plants market. In spite of potential commercial viability of cultivation of medicinal plants, the activity has not widely spread among farmers. So for quality product, improved agro-technique on urgently required for adoption by farmers for cultivation and mass propagation.

UNESCO in its report mentioned the use of traditional medicine and medicinal plants in most developing countries as maintenance of good health. In industrialized country, an increasing reliance on the use of medicinal plants has been traced to the extraction and development of several drugs and chemotherapeutics as well as from traditionally used rural herbal remedies. About 90% of medicinal plant requirement of the Indian herbal drug industry and organic cosmetic manufacturers are met from the wild medicinal plants, collected in a pattern that is not based on the principle of sustainable yield harvesting practice. It is observed that more than 800 species of plants are being currently used by the industries for large scale production of herbal products of which less than 20 species are cultivated commercially. Thus, more than 95% of the medicinal plants used by the Indian industry are collected from the wild conditions (forests). Such raw materials collected from the wild (forests) are almost collected by the forest villagers mostly tribals and other forest dwelling communities.

Constraints

- Poor agricultural practices.
- Poor harvesting (indiscriminate) and post-harvest treatment practices.
- Lack of research on development of high-yielding varieties, domestication etc.
- Poor propagation methods.
- Inefficient processing techniques leading to low yields and poor quality products.
- Poor quality control procedures.
- High energy losses due to processing.
- Lack of current good manufacturing practices.
- Lack of R&D on product and process development.
- Difficulties in marketing.
- Lack of local market for primary processed products.
- Lack of trained personnel and equipments.
- Lack of facilities to fabricate equipment locally.
- Lack of access to latest technologies and market information.

Solutions

- Protect them in their natural environment.
- Scientifically cultivate them in sufficient quantity so that they become much cheaply available in market.
- Develop technologies which provide more income to farmers so that they are encouraged to take up cultivation of M & APs.
- Some agency should be given the responsibility of collection, preservation, processing and value addition and relevant technological processes, market information's to facilitate new entrepreneurs.
- Centres for development of entrepreneurship in tribal areas for postharvest management and value addition should be encouraged.
- R&D organizations and Govt. agencies should provide funds for research projects.
- Technologies developed should be evaluated, modified, adopted and demonstrated.
- Tribals should be trained and encouraged to establish forest produce processing centres for gainful employment and income.
- R&D organizations should examine nutritive values, shelf life, cost economies and energetics involved in production process.
- M&APs should be introduced under Agroforestry systems.

Processing and Value Addition

As markets are not prominent in the state processing and manufacturing units for value addition products based on medicinal plant have not developed in any significant way. Most of the products reach distant markets and initiatives at private or state level has not been prominent.

Profitability

The margin accruing to farmers through trading of medicinal plant produce/collection is uncertain and depends on the volume marketed. The market prior unregulated prices are usually determined through bargaining strengths. Sellers being mostly from lower economic group are hardly in a position to influence the price level.

Poverty reduction through development of M&APs sector

- In Medicinal and Aromatic plants harvesting from the wild, it is crucial that the balance between commercialization and conservation be maintained; this needs to involve local communities and local district or area authorities, as well as national regulatory bodies.
- Widespread education and popular information about natural products as outputs from farms and enterprises.
- Promotion of products and technologies that are suitable for adoption by poor small-farm or rural families and awareness of the good agricultural and processing practices necessary to support their production and market access.
- Influencing the policy, regulatory and enforcement framework covering infrastructure, communications and information provision, technical assistance, business development services, inspection and certification and financial support.

Market Mapping

Market mapping and market information system of medicinal and aromatic plants are highly topical and relevant in view of the emerging income generation potential of medicinal plant sector and the large involvement of rural and tribal community is this segment. The Indian Society of Agribusiness Professionals (ISAR) in 2005, surveyed the market linkages in the medicinal plant trade sector and eventually strengthen economically the back-end producers. SPWD, a New Delhi based organization have give a proposal to promote organic farming of herbal medicinal and aromatic plants in the country, in a large scale.

There are no major markets in the state of Jharkhand. Market information system is nonexistent at the grass root level. However, big farmers who cultivate medicinal plants have access to district markets and are well aware of market trend. The margin accruing to farmers through trading of medicinal plant produce/ collection is uncertain and depends on the volume marketed. The market price is unregulated and prices are usually determined through bargaining strengths. Sellers being mostly from lower economic group are hardly in a position to influence to price level.

Marketing strategies

- Marketing strategies should ensure benefits to primary gatherers, bridging gap between prices realized by the primary gatherers and market prices.
- Present form of contractor system should be replaced by controlled markets with protection to tribal co-operatives/ peoples' institutions/government agencies.
- Role of institution like "gram sabha" should be sensitized for creating better awareness, greater transparency and dissemination of information on prices and markets.
- Healthy competition amongst the peoples' institutions, co-operatives and State agencies should be allowed through suitable people-friendly forest policies.
- Promotional marketing boards should be set up at the district level to disseminate market information.
- An interface between traditional knowledge and modern concepts needs to be forged for M&APs production, marketing and processing with value addition.
- Provision of adequate infrastructure for storage, processing, sale, transportation and skill development programmes may be undertaken to ensure use of sustainable as well as scientific methods of extraction of M&APs.

SWOT analysis of Medicinal Plant Sector in Jharkhand

Strengths

- The state has rich endowment of natural resources in medicinal plants and this would help in income generation of village communities in the area surrounding the forest region.
- Wild collection has been the major income generating source for large group of people in other states (Chhattisgarh, M.P. etc.). This could be replicated in Jharkhand as well.
- The agro-climatic feasibility of the state supports wide number of species and this could develop a large group of medicinal plant cultivators and ensure them higher returns than traditional crops.

- Medicinal plant sector in the state is unstructured. The cultivation of MAP is in experimental stage and so no farmer is interested to take risk.
- Govt. action has not been very prominent to promote farmers to practice medicinal plant cultivation.
- Experience in the collection, cultivation and trading of a variety of products and in expansion.
- Favorable economic situation, climate, natural environment, and good conditions for organic classification.
- Established value chains; large, private company leaders and out-grower and collector interests.
- Availability of skills, technology to add value, improve quality and marketability.
- Pro-poor crops that may be easily integrated into smallholder farming systems.

Weaknesses

- Absence of national policies, regulations, policing and control with respect to collection and cultivation in the wild.
- Small domestic market; difficulty of certification and registration for prime food and drug markets.
- Limited capability among small producers and many value chain actors.
- Absence of farmer collaboration and scarcity of contract production arrangements.

Opportunities

- Buoyant demand for most products as raw materials, or part-processed or finished goods.
- Solid financial returns and enhanced household incomes from collection, production and marketing.
- Scope for the improved management of wild stock and the better extension and conservation of natural resources.
- Non-farm enterprise development and employment creation.

Threats

- Risks involved due to high costs of marketing, and large investments required for modern processing.
- High transaction costs and difficulties faced by small producers in complying with the highest standards in purity and quality and with the conditions in supply and import.
- Absence of product technical guidelines and comparative financial and economic returns.
- Volatility of market prices and fierce and increasing competition, especially in export markets.

Conservation & Developmental strategies

World Commission on "Environment and Development" had recommended strategies for development of M & APs.

- Conducting an item wise resource inventory survey.
- Initiating a conservation programme related to protection, management and propagation.
- Formulating a systematic harvesting / collection programme.
- Conducting action research on collection, extraction, drying, storage, processing and marketing.
- Developing organization for procurement, storage, marketing and trading.
- Conducting socio-economic studies for improving the economy and social status of people by improved utilization, marketing and value addition.
- Training and awareness creation for scientific collection, storage, processing and marketing.
- Feasibility studies on important products to attract investment.

It is not uncommon that the species particularly herbs and grasses yielding products of high demand are heavily collected thereby reducing the distribution and density of such species. There are good number of species which were well distributed earlier is now gradually becoming rare. It is imperative to identify such species and provide protection to avoid the species totally becoming extinct. *In situ* conservation of such species ensures the natural process of evolution to continue without any hindrance. *In situ* conservation efforts should also be made for long term *ex situ* conservation of the gene pool.

Local people enjoying the benefits of forest plants should be trained in their collection using scientific methods. Indiscriminate and unscientific exploitation should be avoided. A conservation package including protection, management harvesting and propagation of the species yielding medicinally important products of subsistence importance should be evolved for each session to ensure the biodiversity conservation. Suitable silvicultural treatments are to play key role in the sustainable management of the resource. The forest policy (1988) had given due importance to peoples participation. Participatory forest management is the new concept for regenerating the degraded forest. The Govt. should aim at the programmers leading to the socio-economic development of the people to bring down their level of dependence on forest. Amelioration of the living condition of such people will certainly help to enrich the forest with simultaneous conservation of biodiversity of the region.

Mushroom Cultivation

Mushroom is a very delicious and important proteinous vegetable crop. It has a great potential in India and its production is exponentially increasing in areas of favorable climate and where the technological back up is available. There are around 38,000 mushroom varieties known to exist in the world and 914 mushroom have been reported from India. However, about 100 of these are considered edible. Of the edible varieties, the most popular ones are the *Agaricus bisporus* (the European or white button mushroom), *Lentinus edodes* (Shiitake or the Japanese mushroom), *Pleurotus* sp like *Pleurotus ostreatus* (American oyster mushroom), *Pleurotus sajor-caju* (Dhingri or Indian oyster mushroom), *Volvariella volvacea* (Paddy straw mushroom), *Auricularia polytricha* (Kanchpra) and *Calocybe indica* (Milky mushroom). It has become popular in northern states viz., Himachal Pradesh, Punjab and Haryana. These states are remarkable producer of White button mushroom (*Agaricus bisporus*) with average yields of 20-22kg per quintal of compost. It is now becoming popular in in Northern Eastern States.

The two most commonly grown species of mushroom in India are white Button mushroom and Oyster mushroom. Most of the production of white button mushroom in India is seasonal. The cultivation is done using conventional methods. Usually, unpasteurized compost is used, hence yields are very low. However, in recent years, yield of mushroom has increased as a result of introduction of improved agronomic practices. Cultivation of the common white button mushroom requires technical skill. Apart from other factors, the system requires humidity, two different temperatures i.e. Temperature for spawn or vegetative growth (Spawn Run): 22-28^{0}C, for reproductive Phase (fruit body formation): 15-18^0, Humidity: 85-95% and enough ventilation during substrates that are sterilized are easily contaminated unless spawned under very aseptic conditions. Therefore steaming at 100^0 C (pasteurization) is more acceptable.

Oyster mushroom(*Pleurotus* spp.) is known as Dhingri in India, which includes many species e.g. *Pleurotus ostreatus, P. sajor-caju, P. florida, P. sapidus, P. flabellatus, P. eryngi*i and many other edible species. Mushroom growing is an occupation requiring perseverance, patience, intelligent observation and a skill that can be developed only through intelligent experience.

Pleurotus mushroom requires a temperature of 20ºC to 30ºC, both for its vegetative growth (spawn run) and reproductive phase, i.e. for formation of fruit bodies. The suitable cultivation period at high altitude - 1100-1500 meters above mean sea level – is March to October, mid altitude - 600-1100 meters above mean sea level – is February to May & September to November and at Low altitude - Below 600 meters above mean sea level – is October to March.

Wild Edible mushroom

In the state of Jharkhand wild grown mushroom have already gained popularity and as such there is a large scale acceptance of this unique vegetable. The area represented by the Ranchi Centre is rich in biodiversity and tropical and subtropical mushroom species are being grown in the area. Several reasons enabled rural folks to participate successfully in the harvest of wild edible mushrooms. Regular survey of markets provide important information on harvesting strategies and management plans implemented to ensure the lasting presence of these socially and economically important species.Introduction of simple and appropriate low cost technology for utilization and production of some of these socio-economically important indigenous species can be undertaken in the area. This becomes even more relevant when Food and Agricultural Organization has recommended the use of edible mushrooms as food supplement for protein deficient populations of developing and underdeveloped countries.

Fig 9.1 Local Khukhri collected from Forest in Ranchi

Wild Edible Mushrooms for Nutritional security

Mushrooms are low in calories, high in fiber, and contain many important vitamins and minerals. Some also have medicinal properties such as complex carbohydrates that strengthen the immune system. Edible mushrooms (*Pleurotus*) are considered a healthy food as it contains protein (around 20-30% by dry weight). This can be useful for vegetarians or anyone looking to increase the protein content in their diet. Fiber, which helps in lowering cholesterol and is important for the digestive system. Niacin and other important B vitamins which as certain B vitamins are found in animal tissue but not plants. This can be another good supplement for vegetarians. Vitamin D-essential for the absorption of calcium. Copper - aids in helping the body absorb oxygen and create red blood cells. Selenium - an antioxidant that helps neutralize free radicals, thus preventing cell damage and reducing the risk of cancer and other diseases. Mushrooms contain more selenium than any other form of produce. Potassium - an extremely important mineral that regulates blood pressure and keeps cells functioning properly. Other important minerals - Such as phosphorous, zinc, and magnesium, low levels of fat, calories, and sodium, no cholesterol, also contribute towards nutritional value of mushroom. The nutritive value of button mushroom has been shown in table2.

Table 9.3 Nutritive value of raw Mushrooms (Italian brown button)

Nutrients	Nutritional value per 100 g (3.5 oz)
Carbohydrates	4.3 g
Fat	0.1 g
Protein	2.5 g
Vitamins	
Thiamine (B_1)	(9%) 0.1 mg
Riboflavin (B_2)	(42%) 0.5 mg
Niacin (B_3)	(25%) 3.8 mg
Pantothenic acid (B_5)	(30%) 1.5 mg
Vitamin C	(0%) 0 mg
Trace minerals	
Calcium	(2%) 18 mg
Phosphorus	(17%) 120 mg
Potassium	(10%) 448 mg
Sodium	(0%) 6 mg
Zinc	(12%) 1.1 mg

Percentages are roughly approximated using US recommendations for adults.
Source: USDA Nutrient Database

In developing countries mushroom cultivation is gaining more attention as a possibility to use agricultural wastes, and to help alleviate poverty and food insecurity.Certain mushroom species are rather easy to grow while others demand more specific cultivation methods and temperature. In general, the life cycle of a crop is rather short (varying from some weeks to a few months). Once the cropping cycle is completed the spent mushroom substrate/compost (SMS) can be used as a soil conditioner, or in some cases as animal feed.Certain mushroom species however, like the Paddy Straw Mushroom and the Button Mushroom can only be cultivated on fermented substrate or compost. Paddy straw is the best substrate for in terms of yield, biological efficiency (BE) of *Pleurotus florida* and blue oyster *Hypsizygus ulmarius* and maximum water use efficiency. Soybean substrate is next better alternative to paddy straw for blue oyster (*Hypsizygus ulmarius)* production. The oyster mushroom requires less water (16 to 38 litres per kg mushroom production).

Process of composting for mushroom cultivation is a more complex process and needs to be promoted and popularized in the plateau region. Lack of information on technical aspects of mushroom production requires describing the complete process of composting of agricultural wastes as well as the specific cultivation methods of each of the appropriate species. The process of obtaining good quality spawn and spawn production requires some sophisticated equipment. Provision of loan or financial support to the farmer will increase the adoption of this technology on mass scale.Harvesting and post-harvest handling of mushroom is another important area that requires special attention.

Diversification through Tuber Crops

Root and tuber crops are the most important food crops after cereals and grain legumes. These groups of crop include cassava, sweet potato, taro, elephant foot yam, yam, yam bean and arrowroot. These crops can play an important role in food and nutritional security. Tuber crops being source of starch can play major role in global food system. Tubers of these crops are used as human food, livestock feed and industry. Tubers of cassava and sweet potato are consumed after boiling and baking. Tubers of these crops serve as subsidiary food. Yams, aroids and Chinese potato are considered, as vegetables and their potential remain largely untapped. Yam bean is a ready-made food crop. Its tubers are consumed as a salad. Fresh and dehydrated tubers serve as good animal feed. Cut and culled tubers can also be used as animal feed. Green tops of tuber crops are also used for animal feeding. Sweet potato vines are excellent green fodder for livestock.

Tuber crops also serve as an industrial raw material. The starch extracted from cassava is used for preparing sago, wafers, and porridge and also used in textile and adhesive industries. Biodegradable plastics are also manufactured utilizing cassava starch. Good quality alcohol is extracted using cassava starches. Tubers of sweet potato are also used for starch and liquid glucose extraction. Good quality alcohol is extracted from cassava tubers. Carotene and colouring agents like anthocyanin pigment also extracted from identified sweet potato varieties. Arrowroot starch is highly priced and mostly used in baby weaning foods and confectionary. Apart from tubers and cormels, plant parts like leaves, petioles and stems are also used for human food and animal feed (Table 9.4).

Table 9.4 Common and botanical name and edible parts of root and tuber crops

Common name	Botanical name	Parts of plant used as food
Cassava	*Manihot esculenta*	Root and leaf
Sweet potato	*Ipomoea batatas*	Root, young vine, leaf and petiole
Greater yam	*Dioscorea alata*	Tuber
White yam	*Dioscorea rotundata*	Tuber
Lesser yam	*Dioscorea esculenta*	Tuber
Elephant foot yam	*Amorphophallus paeniifolius*	Corm and Cormel
Taro	*Colocasia esculenta*	Corn, cormel, leaf
Tannia	*Xanthosoma sagitifolium*	Corm and Cormel
Yam bean	*Pachyrrhizu serosus*	Root
Chinese potato	*Plectranthus rotundifolius*	Tuber
Arrow root	*Maranta arundinaceae*	Rhizome

Nutrition

Tuber crops are rich in minerals and vitamins. Sweet potato rich in â-carotene (8-14 mg/100 g fresh tubers) and contains 24 mg of ascorbic acid per 100 g of fresh tuber. Taro (*Colocasia esculenta*) contains as high as 140 mg of phosphorus per 100 g of tuber and 153 mg of calcium is found in 100 g of coleus tuber. Many tuber crop species are bestowed with immense medicinal value.

Energy content of tubers

These crops have attracted increasing attention because of their remarkable ability to produce tremendous amounts of food, at a modest cost in a short period of time. These crops have the potential to transform a high fraction of biological yield into economic yield/edible energy (Table 9.5).

Table 9.5 Calories produced by the various crops per unit area per unit time

Crop	Calories/ha/day
Cassava	250000
Maize	200000
Rice	176000
Sweet potato	150000
wheat	110000

Adaptability

These crops are adaptable to a wide range of environment and soil. Tuber crops are having great degree of drought tolerance, once it is established. Tuber crops grow well in tropical to subtropical climate with optimal rainfall of 100-150 cm. The agronomic characters presented in the Table 3 indicates its high adaptability to wide range of soil conditions *i.e.* marginal to fertile soil with pH ranging from 5-6.5. Better yields are obtained in well drained sandy loam soils. Cassava and sweet potato can give good return even in poor soil. Water logging is deleterious for tuber crop cultivation except taro. Long storability of tubers under the soil/delayed harvest is another feature, which is highly suited to resource poor rainfed farmers.

Major tuber crops

The most important tropical tuber crops are Cassava (*Manihot esculenta* Crantz), Sweet potato (*Ipomoea batatas* L), Yams (*Dioscorea* spp.), Aroids (*Colocasia* spp.; *Xanthosoma* spp.; *Amorphophallus* spp.), Chinese potato (*Solenostemon rotundifolius*), Arrow root (*Maranta arundinaceae*) and Yam bean (*Pachyrrhizus erosus*).

DIVERSIFICATION THROUGH AGROFORESTRY

Trees provide a range of products such as fruits, fodder, fuel wood and environmental services such as conserving soil and water, carbon sequestration, etc. A number of agroforestry systems have been developed and some of them have been adopted by farmers in semi-arid regions of the country. Some of them are fruit tree based systems *viz.* mango, guava and aonla, timber based systems which are distributed in an area of 13 m ha. Management of trees in conjunction with crops in rainfed areas minimizes the risk associated with stress period through diversified components and through efficient utilization of limited natural resources. Agroforestry is practiced with following major objectives (Korwar *et al.,* 2014).

1. Gainful utilization of off-season precipitation
2. Income stabilization
3. Soil and water conservation
4. Insurance against weather aberrations and
5. Mitigation of climate change

Special conditions/constraints have been identified where agroforestry systems can be successfully taken up Nair (1993). Subsistence farming is one of the many situation where agroforestry can be adopted to stabilize the income of resource poor farmers. Plantation of forestry species on vast extend of degraded lands can help in building healthy soils, reducing soil erosion and improving land productivity. The regions which import fuel, fodder and timber can reduce gain significantly by generating these commodities indigenously. However, certain constraints like land-tenure, low capital, labour requirement may hamper development of agroforestry in the region.

Multipurpose Trees species (MPTs)

Multipurpose Trees species are the species that can provide diverse benefits in various land-use systems. Nair (1993) also defined this as "those trees which are deliberately kept and managed for more than one preferred use, product, and/or service. For example Teak, Mahagony, Gamhar, Bakain, Neem, Arjuna, Siris, Shisham, Kalashisham *etc.* are timber yielding trees which are also called as insurance trees/plants generally grown under different Agroforestry systems and are more accepted by the farmers. There are many other MPTs found throughout the world, among them some are *Acacia nilotica, Albizia lebbeck, Azadirachta indica, Dalbergia sissoo, Gliricidia sepium, Gmelina arborea, Leucaena leucocephala, Moringa oleifera, Pongamia pinnata, Prosopis juliflora, Tamarindus indica, etc.*

Role of agroforestry in management of degraded lands

a. Management strategies for waterlogged and marshy lands: Proper drainage is important to reclaim these areas. Tree species suitable for the waterlogged areas are *Eucalyptus robusta, Syzygium cumini, Salix* spp., *Populus nigra, Terminalia arjuna, Acacia nilotica etc.* In case of marshy areas species like *Baringtonia spp., Bischofia javania, Eucalyptus robusta, E. rudis Lagerstroemia flosreginea, Casurarina equisetifolia etc.* are found more suitable.

b. Management strategies for Sand and desertic lands: Establishment of grasses is must. Trees are to be planted across wind direction. Cultivation of tree species like Teak, *Casuarina* sp., *Acacia tartalis, Prosopis* sp., *Eucalyptus* sp. *etc.* are the most common tress grown under such conditions.

c. Management strategies for degraded land under plantation crops: Plantation of suitable MPTs to the particular area where gaps created can be done for soil binding and water harvesting in root zones as an impediment to the eroded soils. Suitable MPTs like *Eucalyptus camaldulensis*, Acasia, Ailanthus, Casuarina, Prosopis, Neem*etc.* and many economic fruit trees like Guava, Indian gooseberry, Drumstick, Ber, Mango, Cashew *etc.* can be planted successfully.

d. Management practices for Gullied and ravenous lands: Reducing the slopes of the gullies and diverting the surface flow are essential pre-requisites for any successful afforestation programme. Constructing gully plugs (like macro and micro check dams) in the bed of the main and branch gully are playing important role to conserve soil and moisture. Trees species like *Acacia nilotica, Acacia catechu, Azadirachta indica, Albizia lebbeck, Dalbergia sissoo, Prosopis juliflora etc.* are found suitable for these areas.

e. MPTs for salt affected areas: In places where the ground water is saline and scarce in quantity as in arid and semi-arid regions, suitable MPTs are planted. In general, Plantations for fuel wood are rated better for saline soils than timber wood species. Many such species *viz.,Prosophis juliflora, Acacia nilotica, A. catechu, A. auriculiformis, A. arabica, Tamarix articulata, Casuarina equisetifolia, Eucalyptus hybrid, Albizia lebbeck, Azadirachta indica, Terminalia arjuna, Dalbergia sissoo, Leucaena leucocephala etc.* are found suitable for these areas. Fruit trees like Sapota, Guava, Ber, Pomegranate and Custard apple can also be selected for plantation.

f. Management strategies for barren rocky/stony waste/Sheet rock area: In such areas where rainfall is low to medium, many MPTs like *Ecalyptus tereticornis, Melia azaderach, Albizia lebbeck, Acacia catechu, Ailanthus excels, Hardwickia binata, Cassia siamea, Prosopis chilensis, Dalbergia sissoo etc.* can be planted. But for area with high elevation (upto 2000 m), species like *Robinia pseu doacacia, Populus ciliate, Pinus roxburghii etc.* can be selected for planting.

g. Management strategies for steep sloping area: Protective measures can be taken up which includes isolation of the area from man and animal. Vegetative measures like planting of various MPTs *viz., Alnus nepalensis, Betula alnoides, Trewia nudiflora* etc. can be done.

Agroforestry Systems and Practices

Agroforestry systems/practices of India may be classified as traditional and advanced agroforestry systems. Traditional Agroforestry System is characterizedby scattered trees on farm lands/ Parkland systems in which trees are mostly planted on the farm boundaries. In this system farm wood lots or blocks are brought under a particular type of forest species. Plantation of trees on range lands is also categorized in to traditional forestry system. These also includes plantation of vegetative live hedges to be used as live fences for protection of the farms. There is no systematic integration of agriculture and forestry in tradition type of agroforestry systems. On the other hand Improved Agroforestry practiced in rainfed areas systematically integrates Forest Trees and Arable Crops-Based Systems.

There are well characterized and defined agroforestry systems like Alley or Hedge Row Cropping. Fruit Tree-Based Cropping Systems Involving Grasses and Arable Crops have also been developed to tap the potential of synergic effect of these components. Some plantations of forestry species like *Leucaena leucocephala, Tectona grandis, Swetenia mahagony* and Bamboo aimed at getting direct monetary benefits are categorized in to commercial agroforestry systems. There are other agroforestry systems which are used for specific purposes. These include biofuels and bioenergy systems, silvopastoral systems, boundary plantations/ wind breaks and shelter belts, Shade Trees in Plantations.

Table 9.6 Recommended Agroforestry practices for different agroclimatic regions

Agro climatic region	Suggested trees	Crops
Eastern plateau and hill region (Maharashtra, Uttar Pradesh, Orissa and West Bengal)	*Acacia nilotica, Borassus flabellifer, Tectona grandis, Gmelina arborea*	Coarse cereals, pulses,oil seeds, soybean, lentil, mustard
Central plateau and hill region (Madhya Pradesh, Rajasthan and Uttar Pradesh)	*Azadirachta indica, Acacia nilotica, Hardwickia binata, Dalbergia sissoo*	Ground nut, sorghum,pigeonpea, gram,lentil, mustard, soybean, paddy, sesamum
Western plateau and hill region	*Azadirachta indica, Acacia nilotica,*	Ground nut, sorghum,

(Maharashtra, Madhya Pradesh and Rajasthan)	*Hardwickia binata, Dalbergia sissoo, Leucaena leucocephala*	pigeon pea, gram, lentil, mustard, soyabean, paddy, sesamum
Southern plateau and hill region (Andhra Pradesh, Karnataka and Tamil Nadu)	*Tamarindus indica* *Ailanthus excelsa*	Tomato, chili, curry leaf, Cowpea, sesamum, sorghum, pearl millet
	Albizia lebbeck	Cowpea, sesamum, sorghum, pearl millet
East coast plains and hill region (Orissa, Andhra Pradesh, Tamil Nadu and Pondicherry)	*Acacia nilotica, Gmelina arborea, Dalbergia sissoo, Anogeissus acuminata, Prosopis juliflora*	Paddy/pulse, ground nut, sesamum

(Korwar *et al.*, 2014)

Table 9.7 Suitable inter crops for important horticultural crops

Crop	Age	Inter crop
Mango	Up to 7 years	Leguminous vegetables, papaya (filler)
Grapes	Up to 8 months	Snake gourd, bitter gourd on pandal
Apple, Pears	Up to 5 years	Potato, cabbage
Banana	Up to 4 months	Sunhemp, onion
Tapioca	Up to 3 months	
Turmeric	Up to 3 months	
Areca nut	Up to 10 years	Pineapple
Coconut	Up to 3 years	Banana, tapioca, vegetables

(Kumar, 1997)

Benefit–cost ratios of different alternate land use systems

Benefit–cost ratios of different alternate land use systems had been tabulated in Table 3. The Benefit: Cost ratio was reported higher in case of Fruit trees + Arable crops, followed by Forest trees + grasses (2.45). The least was reported in case of arable farming (1.34).

Table 9.8 Benefit–cost ratios of different alternate land use systems (CRIDA, 1993 & 1996)

System	Period (Years)	Benefits: Cost ratio
Arable farming	1	1.34
Forest tree + Sorghum + Pigeon pea	10	1.65
Fruit trees + Arable crops	30	5.53
Forest trees (with Castor intercrops)	10	1.99
Forest trees + grasses	10	2.45

Fruit tree based systems are reported to generate net returns up to Rs. 32,000/ha/year in mango, Rs. 24408/ ha/ year in aonla, Rs. 3916/ ha/ year in guava. Net returns in wood based systems are to the tune of Rs. 21875/ha/year in eucalyptus and Rs. 19695/ ha/year in leucaena which are about three times more that of the annual crops. The benefit cost of these systems can be up to 5.5 and provides reasonable returns during the years of low rainfall(Jasti et al 2014). Integration of livestockand high value intercrops in combination with high density planting, canopy management, organic production practices can improve returns to great extent. The impact of these systems can be further enhanced by integrating them in developmental programs aimed at arresting land degradation and employment generation. Agroforestry systems have tremendous potential to enhance returns to farming communities.

One rational and cost-effective method for building resilience into agricultural systems against the climate change may be agricultural crop diversification. Crop diversification can improve resilience in a variety of ways. Economic incentives encouraging production of a select few crops and the belief that monocultures are more productive than diversified systems have been hindrances in promoting this strategy. However, crop diversification can be implemented in a variety of forms and at a variety of scales, allowing farmers to choose a strategy that both increases resilience and provides economic benefits.

Biofuel Crops

India is a rapidly expanding large economy and faces a formidable challenge to meet its energy needs. India has to face an estimated demand of energy of 1516 million ton oil equivalent (Mtoe) by 2032 which is more than double of the present consumption. Oil provides energy for 95% of transportation needs in our country, whereas it has only 0.4% of world's proven reserve of crude oil. It is important to study the socio-economicand environmental consequences of biofuel crops. Some global studies have favored using marginal or abandoned lands for planting of biofuel crops. This will avoid competition with food crops.Some studies have demonstrated potential of adopting biofuel production crops to createemployment and income opportunities for local populations through contract farming, mainly using: *Jatropha curcas* (ICRISAT, 2007; UNESDA, 2007; Clancy, 2008). Jatropha is claimed to be a hardydrought-tolerant shrub that reclaims the land, prevents erosion, and responds better to organic manure than chemical fertilizers. These properties make this plant suitable to be cultivated both in marginal lands and in small farmers'plots.

Biodiesel

The domestic demand, stagnant production & huge imports make edible oil as unviable option for biodiesel production in India. Biodiesel is a fuel comprised of mono-alkyl ester of long chain fatty acids derived from vegetable oils or animal fats. After processing, combustion properties are very similar to diesel. It is considered a clean fuel since it has no aromatics & almost no sulfur and has 10% built -in-oxygen, which helps it to burn fully.

The demand of energy for maintaining the societal metabolism ofdeveloped and emerging economies like India has led manygovernments to promote biofuels. India started its biofuel initiative in 2003. More advanced and efficient conversion technologies now allow the extraction of biofuels in solid, liquid & gaseous forms from a wide range of biomass sources such as woods, crops & biodegradable plant & animal wastes. In first generation, Biofuels are made from biomass consisting of sugars, starch, vegetable oils, animal fats or biodegradable output wastes using conventional technologies. In second generation, Biofuels are made from lignocellulosic biomass (stalks of wheat, corn, wood) to liquid technology e.g., Biohydrogen, biomethanol and wood diesel. Third generation biofuels are in the nascent stage of development. They are derived from low input/ high output production organisms (algal biomass). Fourth generation biofuels are derived from bioconversion of living organisms (micro-organism and plant) using biotechnological tools.

Blending of biofuels

The Indian National Biofuel Policyaims at blending 20% bioethanol and biodiesel with gasoline and diesel respectively by 2017 (Government of India, 2009). This can be achieved through:

i) Commercially viable lignocellulosic ethanol produced from Agricultural and forestry waste.
ii) An economically cost efficient system available for Algal production and also a commercial scale technology for production of biofuels from Algae either through harvesting and oil transesterification or direct conversion to Bio-oil.
iii) Next generation biofuels from different biomass feedstock.

Jatropha

Jatropha curcas has been found to be one of the most suitable species due to its various favourable attributes like its hardy nature, high oil recovery and quality of oil, etc. The production of seed oil depends on flower production and seed setting. However, in Jatropha, seed yield, which is highly correlated to the number of female flowers, is currently considered as a limiting factor. The ratio of female to male flowers is low in most of the available germplasm. Therefore, increasing the number of female flowers through breeding and management is essential for the success of seed oil production on a large scale.

Recently, Indian scientists have converted coconut oil into biofuel and have approached the union government to commercialize it. Indian researchers have claimed to have run a car on B-20 biodiesel derived from marine micro algae. On the other hand, Nagpur gets India's first ethanol-run bus.

A coherent, consistent and committed policy with long-term vision can sustain India's biofuel efforts. In addition to biofuel, other non-conventional energy harvesting procedures are to be adopted in India like Netherlands and South Korea. Netherlands Organization for Applied Scientific Research developed world's first solar cycle path. Sola Road is a pioneering innovation in the field of energy harvesting. It is a unique concept, which converts sunlight on the road surface into electricity.

Fig 9.2 Biofuel Plantation on marginal and waste lands

Integrated assessments of large-scale biofuel production show that its low Energy Return on Investment compared to fossil fuels, imposes a heavydemand on land, water and labour per net GJ delivered. If biofuels areto replace fossil fuels in a major way in the current economy, the associated land use changes will be significant and will entail tradeoffs across multiple dimensions. The competition forwater with other crops and economic activities will also increase due to their high water footprint. The claimed positive GHG emissions balance will be compromised by the "biofuel carbon debt" of converting forest or shrub ecosystems to energy crops.

Plantation of biofuels plants on the unproductive, degraded and fallow uplands in the eastern plateau and hill region can be help the tribal poor to get some income from these lands. These crop require less care and management for their normal growth and can be better option available for abandoned and permanent fallow lands.

❑❑❑

10

Enterprise Diversification

In the marginal arid and semi-arid agro-ecosystems of India, livelihoods largely depend on crop and animal husbandry and to some extent on non-farm employment. Farming in such areas has been compounded by recurrent rain failure, acute soil erosion, high pest incidences and poor cultivation methods leading to a serious environmental hazard. In contrast to subsistence farming, farmers can opt for entrepreneurial or profit-oriented farming, though without completely abandoning ordinary or traditional farming. This form of livelihood diversification in turn, seems to be having positive implications for income redistribution, employment creation and poverty reduction. This kind of changed scenario calls for marketing innovations with a view not only to understanding them better but also to informing future policy that targets the rural poor.

Diversification

Diversification primarily involves a substitution of one crop or other agricultural product for another, or an increase in the number of enterprises, or activities, carried out by a particular farm. In developing countries such as India, which has been one of the leaders in promoting diversification, the concept is applied both to individual farmers and to different regions, with government programs being aimed at promoting widespread diversification of farm enterprises. Diversification of entrepreneurship is a key factor to sustain small scale farming community in an ever-changing and increasingly complex global economy. There is growing concern about developing entrepreneurship, value•chains and market linkages. Farmer should view farming as more entrepreneurial and reorganize the farming activity to make it more profit oriented.

Farmers as Entrepreneurs

There is gradual change in the components of small scale farming community in India. Farmers are trying new crops and cultivars, better animals, and alternative technologies to increase productivity, diversify production and reduce risk and to increase profits. They have become more market oriented & have learned to take calculated risks to open or create new markets for their products. Smallholder farmers usually do farming mostly for home consumption with rarely any surpluses produced. The produce in excess, if any is taken to the local markets. Producing a commodity exclusively for the market is a rare event.

Opportunities for diversification

Farmers can take up a new enterprises if the income generated by new farm enterprises will be greater than the existing activities, with similar or less risk. Availability of capital, land and labor also affects the decision on entrepreneurships. Opportunity for diversification to an added entrepreneurship is largely governed by the markets for the new products. The United Nations Food and Agriculture Organization (FAO) has been one of the development organizations promoting diversification by small farmers and has produced booklets identifying beekeeping, mushroom farming, milk production, fish ponds and sheep and goats, among others, as diversification possibilities. Other enterprises like aquaculture, sericulture, lac production and duck rearing are some of the potential enterprises in the tribal dominated parts of the country, where natural setup supports lows cost and effective progress of these enterprises.

Lac Production as Enterprise

Lac cultivation is an important source of income for livelihood of the forest and sub-forest dwellers in different states. Besides, it has high potential for generating employment for both men and women in forest and sub-forest areas of Jharkhand, West Bengal, Chhattisgarh, Madhya Pradesh, Maharashtra, Odisha and parts of Uttar Pradesh, Andhra Pradesh, Gujarat and NEH region. It is a high remunerative crop, paying high economic returns to the farmers and also foreign exchange to the country through its export. It forms 20-30% of annual livelihood income of about one million lac growers. India has the largest processing capacity, skilled manpower and traditional knowledge of processing and applications. Presently, India is producing around 20 thousand tons of lac annually and is the world leader in lac production. Lac is mainly produced in India, Indonesia, Thailand, parts of China, Myanmar, Philippines, Vietnam, Cambodia etc. and India is the largest producer of lac in the world.

Lac is a natural resin secreted by an insect *Kerria lacca* (Kerr.), which thrives on the tender twigs of specific host trees *viz., palas* (*Butea monosperma*), *ber* (*Zizyphus mauritiana*), *kusum* (*Schleichera oleosa*), *Flemingiasemialata*, rain tree (*Samanea saman*), *Ficus* spp. etc. The raw lac is the source of three valuable, natural and renewable products *i.e.* resin, dye and wax. *Rangeeni* and *Kusmi* are the two strains of lac insect which are classified based on preference of the insect for specific host plants. *Rangeeni* strain produces two crops in a year known as rainy season *katki* (4 months crop) and winter cum summer *baisakhi* (8 months crop) while, *kusmi* strain also produces two crops in a year namely *aghani* (winter) and *jethwi* (summer), each having a crop duration of six months. The contribution of *rangeeni* lac is about 45 per cent while that of *kusmi* lac is around 55 per cent in the total national production.

Merits of Lac

Lac is nontoxic, biodegradable, eco-friendly and safe natural material for various applications of human contact and consumption. It is a versatile material with wide range of application areas. It possess some unmatched physico-chemical properties for surface coating. Demand for safe natural material for human contact and consumption is on the increase internationally and expected to be sustained.

Lac Insect Systematics

Lac insects constitute a family, the Kerriidae, in the superfamily Coccoidea; order Homoptera of morphologically distinctive scale insects that produce gum like resinous secretions that form a hard scale cover or test (Chamberlin 1923; Varshney 1977, 1984; Lit 2002) is productive ones. The best-known species is *Kerria lacca* (Kerr), renowned for its sticky resinous secretion (lac), which is used commercially as a natural polymer called shellac. The word "lac" is derived from Persian and Hindi words that mean "hundred thousand," indicating the large number of insects required to produce shellac (Encyclopedia Britannica 2007). Cultivation of lac not only provides livelihood to millions of lac growers, but also helps in conserving vast stretches of forests and biodiversity associated with lac insect complex. The lac ecosystem is a complex multi-trophic web of flora and fauna; of the 87 species falling under nine genera recorded from the world, 19 species belonging to two genera, namely *Kerria* and *Paratachardina* are found in India. Ben-Dov (2006) documented 98 species of lac insects worldwide.

Lac cultivation – alternate source of income

Lac is a good source of livelihood of resource poor farmers, tribals inhabiting sub-hilly tracts. It requires no water, meagre fertilizers, labour and pesticides. Lac can be cultivated on the plants established on marginal and degraded lands. Lac production enterprise does not compete with agricultural, horticultural crops for land and other vital resources. It is a highly remunerative enterprise as one hectare of *ber* plantation with *kusmi* lac cultivation can provide a net return of 3-5 lakhs/ha/annum. It is like ancrop insurance specially during drought years as it has good tolerance to adverse climate. A good number of lac host plants like *kusum, palas, ber* etc. naturally occurring in forests and sub-forests are available for commercial exploitation.India has traditional knowledge of its cultivation, processing and utilization besides scientifically documented one.India (eastern plateau) has suitable agro-climate for its cultivation. Lac has high export potential. About 75% of present production is exported.

Sericulture

Rearing of silkworm for silk production is termed as sericulture. The queen of textile fiber, Silk is obtained from mulberry and non-mulberry silkworm. Silk is a synonym of luxury and beauty in textile industry. It is the strongest natural fiber and readily replaces the synthetic fibers (Ude *et al.*, 2014). Apart from textile; it has high technological applications in preparation of biomedical materials and medical implants. India is the second largest producer of silk in the world and produces all commercially used silk variants both mulberry and non-mulberry. Out of four non-mulberry silk varieties viz. Tasar, Oak-tasar, Muga and Eri – the tasar keeps an important position in both for its production and quality of fabrics. Tasar is a superb forest produce and so called a superb Vanya-silk. India has two varieties of tasar cocoons namely- tropical tasar cocoons produced by the silkworm *Antheraea mylitta* D. and the temperate tasar cocoons produced by *Antheraea proylei* J.

The country has the congenial tropical climate, biosphere, natural resources and manpower required for the overall development of tropical tasar into an industry of immense potential. *Antheraea mylitta* Drury (Tropical Tasar) are abundantly distributed in the tropical belt of India. Various eco- types of *Antheraea* species exist in different parts of India. Distribution of these eco-races is mainly based on forest type, soil type and altitude of the area. Tasar culture in India is an ancient traditional practice of tribes and people of weaker section residing in hilly areas. It is practiced by nearly 1.5 lakhs families in different states viz. Jharkhand, Chhattisgarh, Odisha, West Bengal, Bihar, Maharashtra, Madhya Pradesh, Uttar Pradesh, Andhra Pradesh and Telengana. Jharkhand ranks first among the tasar producing states with a contribution of about 60-65% of country's production and Chhattisgarh and Odisha rank the second and third respectively. Tasar silk worm is polyphagus having a number of primary and secondary food plants. Food plants of tasar are abundantly found in these states. Hence, rearing of tasar silkworm for production of cocoons is an option for alternative income of tribal population residing in hilly areas of these states. Some important eco-races of tropical tasar silkworm and their characteristic features are listed below (Table 10.1).

Tasar cocoons have unique identity for their shape, size and quality of yarn. Tasar fabrics are very popular all abroad. The dress material made out of tasar yarns are generally purchased by the rich man of the society. Traditionally, attire of tasar silk is worn in festivals and holy occasions in India but as a whole affluent people of the society wear it as a status symbol all around. As such, money flows from rich to poorest of the poor of the society in tasar culture.

Generally tribals of tropical belt depend upon the forest produce. They earn their livelihood by selling items such as seasonal fruits, seasonal vegetables, medicinal fruits, roots and leaves, oil seeds of different kind, mahua, honey, lac and tasar etc. They also rear goat, sheep, pig, hen and other domestic animals for earnings. Tasar culture is a popular and most important avocation of tribal people residing in forest and hilly areas. Tasar gives them additional income for their livelihood in a very short span of time. Depending upon the season, they can earn 30,000 to 50,000 rupees within a period of one to two months. The potential rearers who have sufficient number of tasar food plants can earn up to one lakh in a year from tasar culture. In this avocation, people become benefitted with very less investments. On the contrary, people of forest and nearby forest area don't have

Table 10.1 Characteristics of important eco-races of tropical Tasar silkworm

S. No.	Eco-type	State where predominantly produced	Voltinism	Single cocoon weight (g)	Shell wt (g)	Filament length (m)	Average Denier	Recovery of silk in gram /1000 cocoons (g)
1	Daba	Jharkhand & Bihar	Bivoltine & Trivoltine	12-14	1.45-2.0	850-1200	10	750-1000
2	Modal	Odisha	Univoltine	15-18	3-4	1400-1600	10-11.5	1800
3	Raily	Chhattisgarh & Madhya Pradesh	Univoltine/ Trivoltine	15-18	2.5-3.0	1400	10	1500
4	Sukinda	Odisha	Trivoltine	12	1.00-1.45	850	10	750-1000
5	Bogai	Odisha	Bivoltine & Trivoltine	10	1.3-1.4	700	8	750
6	Nalia&Lari a	Odisha & Jharkhand	Bivoltine	7-9	0.80-1.0	550-650	7-8	550
7	Tira	West- Bengal	Bivoltine	7-9	0.8-1.0	550-650	7-8	550
8	Sarihan	Jharkhand & Bihar	Trivoltine	8	0.7-0.8	570	8	440
9	Bhandara	Maharashtra	Trivoltine	8.5	0.8-0.9	670	7	450
10	Andhra Local	Andhra Pradesh &Telengana	Bivoltine	8.0	0.9-1.0	600-650	8	550

sufficient agricultural land to cultivate. Moreover, topography of the area is such that the rain water drains out from slope land of the hills. As a result of which, agricultural crop viz. paddy, sugarcane and wheat are not feasible for such type of land. Tasar culture is a boon to dwellers of such area which generate income and prevent deforestation. A comparative data on income of tribal people from different sources have been shown below (Table-2).

Table 10.2 Income sources and total income of tribal families

Avocation	Investments (In Rs.)	Average production/ farmer	Sale price of product (In Rs.)	Net income (In Rs.)	Duration of crop (In Days)
Tasar culture	1200 (1st season)	10000-12000 cocoons	20000-24000	18800-22800	30
	1200 (2nd season)	12000-14000	30000-35000	28800-33800	50
Total	2400	22000-26000	50000-59000	47600-56600	80
Agriculture	25000	5000 Kg (grains)	50000	25000	160
Poultry	1000	20	5000	4000	120
Animal husbandry	10000	5	25000	15000	180
Collection of forest produce	Nil	-	15000	15000	365 (seasonal once in a year)
Others	Nil	-	5000	5000	365 (seasonal)

Rearing of tasar silkworm is done by the member of any age group (above 10 yrs) of the family and arrangement of time for watch and ward duty can be managed as per their convenience. Tasar culture is also done by woman farmer as a part time job. Thus, it provides an additional income. Popularization of this tasar culture is beneficial to mankind. It helps to save the environment and maintains ecological balance. Tasar culture is a part and parcel of sericulture and it has ample opportunity of employment generation for tribal population.

Seed Production

The Eastern India region including Jharkhand, Bihar, Odisha, Chhattisgarh, West Bengal, Assam and Eastern U.P. has the potential to usher in the second Green Revolution in the country. This will be realized if the challenges in the form of low productivity, erratic climatic conditions, inadequate support and infrastructural facilities are addressed. The growing need for food and the flattening yield levels in Punjab and Haryana while posing a problem, is at the same time an opportunity. Rice is the major crop in the region along with other important crops like maize, pulses, oilseeds & wheat. While initiatives are underway for enhancing productivity in different agro-ecological sub regions, the challenges are daunting. The effective way forward would be to constantly explore newer options to meet the twin objectives of enhancing production & productivity. There must be special emphasis on technologies that would benefit the resource poor and marginal farmers, especially from the tribal populations.

Jharkhand was formed in November 2000 after figuring 79 lakh hectares ecological area from Bihar state for enhanced execution and planning of expansion activities of an economically different zone. Agriculture is the chief stay for 80 percent of rural population of Jharkhand. Agriculture is their source of employment and main income engendering activity. The agricultural economy of Jharkhand is differentiated by low productivity, dependence on nature, low investment, mono-cropping with paddy as the overriding crop. The cultivable land resource of the state has good potential for higher production of horticulture and forest products. The forest offers adequate biomass to feed its soil. However, Soil erosion and crash to reprocess the biomass is diminishing the fertility of the soil. Therefore, a judicious soil, land and water management is necessary that can perk up agriculture efficiency. Jharkhand is gifted with rich and vast natural resources primarily those of forest and minerals, 80% of its population inhabiting in 32620 villages and rely principally on agriculture and associated activities for their source of revenue. In agricultural segment there is extent for fetching supplementary area under cultivation in the course of horizontal and vertical expansions, expanding area under irrigation, increasing production and productivity of food crops through most efficient use of inputs like fertilizers, seeds, agricultural tools, implements, pesticides etc. It is hoped that the synergy generated by the efforts of all stakeholders would be a harbinger leading to great benefits for the state of Jharkhand.

Table 10.3 A statistical panorama of Jharkhand agriculture

Particular	Area in %	Area (Lakhs Ha)
Total geographical area		79.71
Total Cultivable Land		38
Net Sown Area	28.08%	25.75
Current Fallow	11.12%	8.87
Other Fallow	8.46%	6.75
Forest	29.20%	23.28
Barren Lands	7.20%	5.74
Non – Agricultural Use	8.60%	6.86
Pasture and Other Grazing Land	2.48%	1.97
Cultivable Waste Land	3.44%	2.74
Irrigated land	12.73%	3.007
Cropping Intensity	125%	

Agro Climate Zones

In general, the plateau region of Chhotanagpur and Santhal Parganas is characterized by humid and sub humid tropical monsoon in sub zone IV, sub humid to subtropical in sub zone V and humid to sub-tropical in sub zone VI, undulating Plateau, hills and mountains, non-existence of perennial rivers, initially high forest cover, which is depleting fast due to mineral and industrial exploitation and encroachment, concentration of tribal population in many areas with a life style geared to forest ecology, pockets with chronic drought conditions, erratic rainfall, low ground water level, high soil erosion, insufficient agricultural dairy and fishery activities, rising agricultural unemployment and acute poverty. Important characteristics of the three sub zones are:

Sub Zone IV

1. Low water retentive capacity of the soil particularly that of uplands.
2. Late arrival and early cessation of monsoon and erratic and uneven distribution of rainfall.
3. Lack of safe disposal of runoff water during monsoon and water storage and moisture conservation practices for raising Rabi crops.

4. Drying of tanks and wells by February results in no Rabi crop production.

Sub Zone V

1. Late arrival and early cessation of monsoon.
2. Erratic/uneven distribution of rainfall.
3. Low water retentive capacity of soils.
4. Lack of soil and water conservation practices.

Sub Zone VI

1. Uneven distribution of rainfall.
2. Low water holding capacity.
3. Eroded soils.
4. Poor soil fertility

Seed Availability

Analysis shows the requirement of 46 lakh tone of food grains for a population of 26 million against the current production of 22 lakh tonnes. The short fall in food grain production is highest in sub zone IV (Central north eastern plateau) followed by that in western and southeastern plateau zone.

Last season (2014) witnessed a deficit of about 13500 quintals in supply of wheat seeds alone. It would have been larger, had there not been a last minute arrangement of large quantities of PBW 343, HUW 234 and UP 262 from various agencies. Similarly, barley, a crop more suited for rain fed conditions, also faced a shortfall of 160 quintals in terms of quality seed. This could be checked at this quantity thanks to the arrangement of 2000 quintals of seeds by HIL. In case of pulses the deficit was nearly 19000 quintals and that in case of oilseeds it was to the tune of 3000 quintals. The short supply has been faced on a similar scale in case of maize as well as millets. Significance of seeds for bringing a good harvest home cannot be overemphasized. How targets of food production can be achieved when we are short of seeds, it is a gaping question?

Strengthening the Seed Supply System

There is an apparent need to step up investment in both public & private sector. The prevailing seed replacement rate which is less than 10% in case of rice and wheat cannot sustain higher productivity growth rate. Therefore the objective of the seed production programme involves increasing the seed replacement rate, promotion of hybrid seeds and crop varieties, which are new and adapted to the agro-climatic conditions of the state. Seed production on Government farms may be revived at the same time the government farms need to be equipped with irrigation, storage and processing facilities. The state owned seed corporation can be encouraged to undertake seed multiplication for public sector varieties released by ICAR-RCER, RC, Ranchi and B.A.U. Ranchi to make them timely available to the farmers at reasonable cost. Farmers' seed co-operatives may be formed and be encouraged to take up production and supply of quality seed to the farmers. While the seed production under public sector has to be revived, seed production through farmers' direct participation under seed-village program may in long term turn out to be the corner stone of the strategy in coming decades. State seed corporation ICAR-RCER and B.A.U. have an increased role to play by developing an extensive scheme of capacity building of farmers in seed production of open pollinated as well as hybrid varieties of crops with an arrangement to buy-back seeds produced under seed village and provide processing and marketing support. Private seed companies should also be roped in to play larger and positive role in ensuring availability of latest technology to the farmers and also to ensure fair competition with public sector seed companies for the benefit of the farmers.

Based on area under the important crops and seed multiplication ratio, an estimate of areas for each crop to be earmarked for seed multiplication is being presented in table given below.

Table 10.4 Estimate of area under important crops in Jharkhand and area required to be earmarked in order to meet entire seed requirement

Crop Group	Crop	Total Area Sown * Area in 000 ha	Seed Requirement** (Quantity in Q)	Area to be ear marked for seed production (000 ha)
Cereal	Paddy	1379.54	185125	17.24
	Wheat	200	89100	10.00
Millets	Maize	251	6000	2.51
Pulses	Bengal Gram	200	34425	13.33
	Black Gram	93.07	4700	2.33
	Green Gram	24.82		0.62
	Horse Gram	6.2	4747	0.16
	Lentil	60	15750	2.00
	Pigeon Pea	186.13	24015	1.86
Oilseeds	Indian Rapeseed	200	6260	2.00
	Niger	10	325	0.10

*on the basis of *Kharif* 2014 actual and Rabi 2014-15 Targeted, **according to Seed Rolling Plan

Bottleneck in Enterprise Diversification

It cannot be claimed that every enterprise will be successful at all the places as progress of an enterprise depends on available resources, management and most importantly the environment. There are certain barriers on which farmers do not have control cover. This depends on policies and the level of investment in agriculture which can create an environment that encourages profitable, market-oriented farm businesses. Major bottlenecks in diversification on an enterprise are;

Poor infrastructure: Farm infrastructure is basic to start any farm enterprise. Poor roads leading to markets, inadequate storage and market facilities, and even irregular supplies of electricity create very real and practical barriers to developing farm businesses.

Legal framework: Land tenure and ownership issues, farm inputs availability, access to credits are some of the more common barriers that limit the development of a farm businesses. The ability to buy, sell and hire land, the legal status of women and the extent of bureaucratic procedures all affect the new farm businesses.

Social issues: The fear of failure and belief that particular enterprise belongs to particular cast are some of the social barriers that limit the scope of enterprise diversification. Women in business are often not supported or are even discouraged. Extension workers will need to be aware of these social barriers and help farmers deal with them.

Lack of training: Some of the enterprise require high level of expertise to operate them. Training facilities and support must be easily available to farmers. This calls for building of new institutions to provide education and training that can accomplish the right balance of technical knowledge and practical skills.

Limited market linkage: Linking of production to market is the key to success of any farm enterprise. In many of the Indian villages, access to markets is often very limited. Poor communication, quality road, effective and efficient transport facilities, lack of information on product demand often hamper the progress of an enterprise.

❑❑❑

11

Integrated Farming Systems

Small and marginal farmers with their limited land holding and less diversified farming entrepreneurships are prone to the risks emanating from climate, water shortages, falling market prices and can hardly sustain their livelihood. Taking improved and low cost technologies at the door-step of the farmers is the need of the hour and many efforts have been made by the several Government and Non-government organizations, private companies and other supporting groups to popularize the resource efficient technologies in the farmers' fields. Apart from the advantages and benefits that a new technology offers, agro-climatic and socio-economic aspects play an important role in adoption of any advanced technology. Farmers look for the technology options that can increase their income while keeping risk within reasonable bounds and keeping view their own circumstances and level of technical and management skills required in the technology. This calls for farmer's participation, systemic research and extension efforts to be made. In this regard, many programs and schemes have been started to integrate the desirable features of farming system research (FSR) strategy into the mainstream agricultural research so that the technologies developed are relevant, client- oriented and location - specific.

Challenges and Opportunities

The challenges ahead in agriculture sector consist of development of (i) sustainable agriculture, (ii) food and nutritional security, (iii) climatic vulnerability and its impact on crop productivity, (iv) generation and transfer of technology, (v) enhancement in resource use efficiency and water productivity (vi) provision of incentives for agriculture, (vii) promotion of investments in agriculture, (viii) strengthening of institutional infrastructure, (ix) better risk management, (x) corporate agriculture, (xi) peri-urban agriculture, and (xii) tapping the potential of eastern region and introduction of management reforms.

In order to meet out aforementioned challenges, technology integration, scaling up and framing of demand driven productivity enhancing research agenda in a network mode, using both conventional and frontier technologies ensuring scientific management of natural resources and production sustainability is essentially required since agriculture is, by and large, diverse and risk prone. This calls for a switch over from traditional farming to Integrated farming system mode. Some of the opportunities which needs special attention towards sustainable production are - System Mode Production, Intensive Integrated Farming Systems (IIFS): a sustainable alternative for the benefit of small scale farmers, Synergistic role of livestock in farming systems, Integrated fish farming, Conservation and sustainable use of Wetlands, Agroforestry interventions, Conservation agriculture, Harnessing green energy etc. and these all opportunities can be carried out together under one umbrella of Integrated farming system (IFS).

Farming System – The Concept

Farming system approach is a whole farm management system which aims utilize synergetic interdependencies between the different components of the agricultural entrepreneurships and how these components interact with each other in respect of physical, biological and socioeconomic factors not under the household's control. Integrated Farming System delivers more sustainable agriculture. It is a dynamic approach which can be applied to any farming system around the world. It envisages the integration of agro forestry, horticulture, dairy, sheep and goat rearing, fishery, poultry, pigeon, biogas, mushroom, sericulture, lac culture, etc. on need based and by-product utilization of crops with the main goal of increasing the income and standard of living of small and marginal farmers.

Integrating systems is all about bringing crops and livestock into an interactive relationship with the expectation that together they will generate positive effects on outcomes of interest, such as profitability, overall productivity and conservation of non-renewable resources. It involves management practices, goals and lifestyles of humans, social constraints, economic opportunities, marketing strategies and externalities including energy supplies and costs and impacts of farm policies. Integrated farming systems also reflect utilization of natural resources available and their impact which lead towards sustainable agriculture by integrating different components for holistic approach for nutritional, economic and employment security for the farming community as per the need based agro climatic condition.

Aims of integrated farming system

The integrated farming systems aims at improved livelihood, energy and resource efficient technologies, clean environment, balanced nutrition, resource recycling and improving standard of living of the farmers. Besides this, the specific aim of integrated farming systems are:

- It enables to identify the major constraints in increasing farm productivity.
- It enables to provide technological intervention options for improving farming systems at a given resource base through farmer participatory research.
- To understand the physical and socio-economic environment within which agricultural production takes place.
- To gain an understand of the farmer in terms of his/her skills, constraints, preferences and aspirations.
- To comprehend and evaluate existing important farming systems, in particular, the practice and performance of these systems.

Scope of integrated farming system

Integrated farming system is an assembly of synergic enterprises. The scope of system lies in effective integration of other enterprises with crop production activity which depends on many factors such as soil and climate, availability of the resources, land, labour and capital. The present level of utilization of available resources and economics of proposed integrated farming system is an important factor in deciding the entrepreneurial components of the system. Further, the technical and managerial skill_of the farmer govern the choice of enterprises and will play an important role in recycling the products and waste materials of one component and reduce the cost of production.

Advantages of integrated farming system

Integrated Farming System has many other advantages over the traditional farming and thus able to play a vital role in Agricultural sustainability and nutritional security in many ways:

Productivity and Profitability: Integrated Farming System provides opportunity to increases economic yield per unit area per unit time by virtue of intensification of crop and allied enterprises. The system as a whole provides opportunity to make use of the produce/ waste material of one component as input on another component at the least cost.

Potentiality /Sustainability: One of the objectives of IFS is to reduce the chemical use in agriculture for sustaining the agriculture and ecosystem as a whole. In IFS organic supplementation through effective utilization of by-products of linked components as manures is possible and thus it will certainly provide an opportunity to sustain the potentially of the production base *viz.*, soil for much longer periods.

Balanced food/nutritional supplements: IFS is a better platform to ensure nutritional security of poor farmers. Different components of varied nature are linked enabling to produce different sources of nutrition *viz.*, protein, carbohydrate, fat, minerals, and vitamins *etc.*, from the same unit area.

Pollution free environment: Use of chemicals like pesticide, fungicide and herbicide and fertilizer is one of the major contributors to environment pollution. In IFS, the waste materials are effectively utilized by linking appropriate components and thus utilizing the by-product as organic manures. Similarly, the bio control measures help in preventing pollution of soil, water and atmosphere.

Recycling: IFS establish its stability due to effective recycling of produces/waste materials of any one of the component as input on the other component linked in the programme. Recycling their own material at the farm level, the farmer could reduce the cost of production which enables ultimately to increase the net income of the farm as a whole.

Round the year income: Unlike conventional crop activity which provides income at the end of the season, IFS provides round the year income through the sale of eggs, milk, mushroom, honey, cocoons of silkworm, *etc.*

Adoption of new technology: Increased income from round the year production activities of IFS gives a sort of inducement to the small and marginal farmers to go for the adoption of improved technologies.

Solve energy and fodder, fuel and timber crisis: Growing of perennial fodder legume trees in the borders and water courses is a recommended practice in IFS. This practice not only helps in supplementing legume fodder but also enriches soil nutrient by fixing the atmospheric nitrogen. In the cropped land also IFS envisages intensification of cropping by including legume fodder like cowpea either as second tier or as third tier in the system. Integration of agro-forestry provides fuel and industrial wood without detrimental effect on crop activity in the field level.

Employment generation and improved literacy rate: Combining crop enterprises increases the labor requirements and can help in solving the problem of underemployment to a great extent. IFS provide enough scope to employ family labour round the year.

Provides opportunity for agri - based industries: Once the produce of different components linked in IFS are increased to commercial level and if there is glut in the market, leads to the development of allied industries for preserving the by-products.

Improves the standard of living of farmers: IFS improves the standard of living of farming community andimparts a confidence to the farmers that they are like other business professionals. This can further boost the agricultural profession without any reluctance.

Farming system varies from zone to zone, region to region, location to location and even farmer to farmer and needs certain modifications in the developed /modified or/prevailing farming system based on location specificity, farmers' preferences and land suitability.

Choice of enterprises and mitigating the risk

The choice of enterprise in integrated farming system plays also a key factor. The basic points that are to be considered while choosing appropriate enterprise in integrated farming system are soil and climatic features of an area/locality, resource availability with the farmer, present level of utilization of resources return/income from the existing farming system, economics of proposed integrated farming system, farmer's managerial skill and social customs prevailing in the locality. With a view to mitigate risks and uncertainties of income from crop enterprises and to reduce the time lag between investment and returns it is essential that farmers include multi-enterprises in their production programme that yield regular and evenly distributed income throughout the year and are not subjected to vagaries of nature. The choice of components in a farming system must ensure that interaction among the components is complementary with least competitiveness. In integrated farming system, the expected outcome should reveal enhanced income, employment generation, nutrient recycling and alternate land use options. The integrated farming system having agri-silvicultural system having lesser risk, agri-horticultural system having higher income, silvi/horti-pastural system, improved sustainability, solution to energy, fodder, fuel and timber crisis and avoid degradation of forests or land.

Size of integrated farming system

The size of farming system depends upon the choice and size of farming system which varies from place to place depending on the agro-climatic conditions, suitability of the enterprise to the location, availability of inputs and marketing facilities. The system approach, careful exploitation of agro-biodiversity involving crops and species requires for proper identification of varieties and other management practices including processing and value addition to fully harness the potential should also be taken in account. The diversification of crops with the use of possible biodiversity will not only enhance the returns but also bring in sustainability besides preserving valuable biota as compared to monocropping.

Constraints in Integrated farming system

There are certain constraints like heavy investment in the initial stage, especially for the procurement of enterprises, involvement of multi-disciplinary activities likes animal husbandry, fishery, sericulture, horticulture, forestry, agricultural engineering, etc. The non-availability of improved cultivars/ varieties/ breeds of livestock at farms site, lack of know-how especially on the constituents of feed and the possibility of supplementing from their own produces with cheaper rate and lack of marketing for the produces from different enterprises at village level are anticipated in the progress of this technology. In integrated farming system, there is a need to study the sustainability of the identified systems under different topographical situations in the long

Table 11.1 Prevailing and potential farming system in different Agro-climatic zones

S. No.	Agro-climatic Regions	Prevailing Farming System	Potential Farming System	Total area (000, ha) under F.S.
1.	Western Himalayan Region	- Crop based farming system - Horticulture based farming system	- Dairy based farming system - Vegetable based farming system	106.8 (10%)
2.	Eastern Himalayan Region	Rice- rice- rice	Rice + Fish + Horticulture + Vegetable	59.04 (3 %)
3.	Lower Gangetic Plain Region	Jute-rice-mustard	Crops + Horticulture + Fish + Dairy + Duck	108.03 (5%)
4.	Middle Gangetic Plain Region	Rice -Wheat	Crop + Dairy + Fishery + Mushroom	146.43 (3%)
5.	Upper Gangetic Plain Region	Crop + Dairy	Crop + Dairy + Fishery + Bee Keeping + Horticulture	283.05 (5%)
6.	Trans Gangetic Plain Region	Rice/ sorghum-wheat + Dairy	- Fishery + piggery - Crop + Dairy (cross bred) - Crop + Dairy + Vegetable - Crop + Dairy + Poultry - Crop + Dairy + Fishery + Piggery	988.56 (18%)
7.	Eastern Plateau & hill Region	- Crops + Horticulture Crops + Fish/Poultry + Dairy	- Crops + Horticulture +Dairy Crops + Dairy + Poultry + Fishery + horticulture + Duck	309.10 (5%)

contd...

8.	Central Plateau & hill Region	- Crops + Livestock -Maize-Potato- Wheat -Soybean- maize-wheat	- Crops + Livestock + Fishery - Crops + Dairy + Vegetables - Crops + Dairy	972.29 (11%)
9.	Western Plateau & hill Region	- Crop based	- Crops + Dairy + Sericulture - Crops + Dairy - Crops + Dairy + Poultry	816.28 (%)
10.	Southern Plateau & hill Region	- Crops alone - Rice/ maize based cropping system	- Crops + Orchard + Goat - Crops + Poultry + Fishery - Crops + Fishery + Pigeon - Crops + Fishery + Goat - Crops + Fishery + Poultry + Mushroom	480.40 (11%)
11.	East coast plain & Ghat Region	- Rice-rice-fallow-pulses	- Rice-rice-rice-cotton+ maize and duck cum fish - Rice-rice-rice-cotton+ maize and Poultry cum fish - Crops + Poultry Fish + Horticulture + Coconut + Poultry	400.77 (9%)
12.	Western plain & hill Region	Rice- rice - Rice + Fishery -Cashew/ Coconut + Rice	- Rice + fish - Rice+ fish + Poultry - Rice – fish (in rotation)	86.21 (11%)
13.	Gujarat plain & hill Region	Crops + dairy (Local breed)	- Crops + dairy (Cross bred) - Crops + Poultry	345.88 (13 %)
14.	Western dry Region	- Crops, cereals and millets - Crops, millets and pulses	- Crops + Dairy + Horticulture - Agri-Pasture System	333.84 (6%)
	Total			5361.99 (8.36%)

run including high value crops. There is a further need to study the nutrient dynamics of soil with continuous cropping and recycling of manurial resources with different systems over time. The modeling of the identified farming system options is to suit a given agro-climatic and socio-economic situation. There is need to identify the constraints in adoption of identified farming systems by the farmers for further refinement.

The development or refinement of research methodologies and analytical tools suited for integrated farming system research and analysis using on-farm databases is mandatory now a days. The participatory technology development including varietal improvement, on farm value addition through vertical integration and post-harvest processing is essential. The creation of databases and their documentation of different integrated farming system for different agro-climatic conditions and their integration in terms of by product utilization, value addition, post-harvest management and marketing are necessary. The cataloguing and utilization of indigenous technical knowledge concepts in the development of farming system modules should also be kept in mind. The soil health aspects and nutrient flow in farming system or farming system development for the poorest and landless based on common property resource management. The whole farm analysis for household nutritional and income security, and gainful employment for women folk and rural youth plays an important role for success of integrated farming system.

IFS models for dry Regions

- Crop+dairy+biogas+silviculture
- Crop+dairy+biogas+mushroom+fish
- Crop + goat
- Crop+tree+goat
- Cropping + fish + poultry
- Cropping+duck+fish culture
- Cropping+milch animals
- Crop+dairy+goat

Horticultural Interventions as IFS component

Demonstration of improved horticultural technologies to enhance annual household income and providing sustainable livelihood security to the rural people will help the tribal and resource poor farmers to understand the potential benefits of these technologies. This calls for motivation and capacity building of farmers to take up these interventions. Training of the farmers for specific technologies both at their farms as well as in research institutions is an essential component. Demonstration of following technologies in the disadvantaged districts of eastern plateau and hill region was carried out under NAIP.

- Rain water harvesting, conservation and utilization can be the top priorities to enhance productivity of horticultural crops in medium and uplands. The UV stabilized polythene lined *Jalkundas (Dobas)* (3.0mx1.5mx1.0m) for rain water harvesting in uplands and providing life saving irrigation to fruit plants in multi-tier system at farmers' fields offer the requisite advantage in the region.
- Drip irrigation through earthen pitchers in the root zone of each fruit plant can be used in multi-tier system.
- Mulching the root zone of the fruit plants with dry grasses, paddy straw and other organic materials helps in reducing the evaporation loss and conserving the soil moisture.
- The technologies of off-season cultivation of cucurbits in winter under low poly tunnel for early summer harvest and better remuneration
- Cultivation of oyster mushroom and value addition of horticultural produce as income option for farm women.
- Raising healthy seedlings in portrays inside low cost poly house as regular income generation activity,
- Fruit based multi-tier cropping system and vegetable cultivation including crop diversification were demonstrated at farmers' fields in Dumka and Jamtara.

Adoption of different horticultural technologies improved the income per farm family. The benefits of the technologies demonstrated have been described below:

Low poly tunnel cultivation of cucurbits in winter for early summer harvest and better remuneration: Low poly tunnel cultivation of cucurbits in winter for early summerharvest and better remuneration was demonstrated in 88 farmers' fields. Farmers have earned the maximum net income of Rs.1571/- (estimated Rs.392750/ha) from bottle gourd cultivation in 1.0 decimal (40 m^2) area. The crop was sown on 13.12.2012 and the first harvest started on 10.3.13 and continued up to14.5.2013. Total harvest was 191 kg (estimated 47.75 t/ha) which was sold @ Rs.10-8 per kg.

Demonstration of vegetable cultivation: Introduction of new crops like pointed gourd, a perennial cucurbit, was found to be very remunerative. Farmers harvested 163 kg produce from 2.75 decimal (110 m^2) area of pointed gourd planted on 28.02.10 and earned Rs.3347/- during harvest period of six months. The farmers expanded their area under pointed gourd cultivation through propagation of pointed gourd plants by vine cutting.The alternate business of selling rooted plants of Pointed gourd to other farmers which further added to the income.

Vegetable seed production: Training and capacity building of farmers on the technical aspects of vegetable seed production at ICAR RCER Research Centre, Ranchi, farmers have started production of pure seeds of improved varieties of vegetable crops in their field. Farmers are able to produce and sell the pure seeds of brinjal, bottle gourd and cucumber in local village markets (*Hats*). This highlights the fact that vegetable seed production can be a lucrative enterprise in the integrated farming system models.

Income generation from Poly house: Seedling production under small poly house structure is another very important component of IFS that offers the duel advantage of getting income and meeting the quality seedling requirement of the farmers. Farmers can earn income of Rs.43165/yr through raising of seedlings of vegetables, papaya and forest plants in portrays/poly bags inside naturally ventilated small poly house of 64m^2 area. A poly house of 64 m^2 floor area costs about Rs.50000/- and a cumulative income of Rs.101895/- can be obtained in three years through the sale of seedlings. The income generated is much higher than the cost of construction of the poly house.

Demonstration of fruit based multitier cropping system: Fruit based multi-tier system can be opted in uplands and medium uplands. These systems help in climate proofing of agriculture to certain extent and help in diet diversification and meeting the nutritional requirements.An income of Rs.24950/- from filler crop (guava) and intercrops (vegetables like potato, tomato, brinjal, radish, okra and chilli) can be obtained in the 5th year of establishment of fruit based multi-tier cropping system.

Women empowerment: Women were trained for mushroom production and value addition in fruits and vegetables. Several farm women are taking up mushroom cultivation as profitable venture. A bag of mushroom typically yields 1-2 kg of mushroom which fetches a price of Rs.80-120 per kg. From the earnings from this enterprise, farm women contribute in the SHG fund. Compared to paddy straw, cultivation of mushroom on wheat straw resulted in increased production. Twelve women of a SHG produced 678.85 kg oyster mushroom from 446 inoculated bags which was sold at the rate of 120-130 Rs/kg providing a total income of Rs.83146/-. The experience in mushroom production imparted the expertise and confidence to the tribal ladies such that they are now performing the role of expert trainer and are actively engaged in conducting trainings. They act as resource persons in the trainings organized by Department of Soil Conservation, Grameen Banks, etc. and get an additional income in the form of honorarium. The farm women were also trained in preparation of mushroom pickle, tomato sauce and jack fruit pickle and papad making, which is sold in local village markets and line hotels.

Integrated farming system approach advocates a change in the farming techniques for maximum production in the cropping pattern and takes care of optimal utilization of resources. The farm wastes are better recycled for productive purpose in the integrated system. Farm as a unit is to be considered and planned for effective integration of all enterprises. The future of Indian agriculture depends heavily on the development of appropriate farming system as it fits to resource poor farm families of different agro ecological zones. The endowment of abundant sunshine, long growing season, responsive soil type and combination of surface water, ground water and seasonal rains and above all a progressive peasantry offer vast scope for an intensive farming system through multiple cropping and diversified farming including animal husbandry, forestry, sericulture, fisheries and the like. The challenge is to upgrade technological and social disciplines on a continuous basis and integrate these disciplines to suit the region and the farm families in a manner that may ensure increased production with stability, ecological sustainability, nutritional security and equitability.

Efficient utilization of scarce and costly resources is the need of hour to make crop production a profitable venture in the present competitive scenario. Concept of integrated farming systems which assembles and ensures synergy in agriculture and allied agro-enterprises is the best option to achieve this goal. Residue recycling, input-output relations among different enterprises, improved water and fertilizer use efficiency and sustained income with clean environment are the prominent features of an IFS. It ensures efficient and judicious utilization of scarce resource and brings dividends in the long run by enhancing water productivity leading to the national goal of "per drop more crop". An integrated farming system approach with better planning and utilization of the available resources will bring bright prospects for the farm economy as a whole.

□□□

12

Conservation Agriculture

Attaining food security for a growing population and alleviating poverty while sustaining agricultural systems under the current scenario of depleting natural resources, negative impacts of climatic variability, spiraling cost of inputs and volatile food prices are the major challenges before most of the Asian countries (Bhan and Behra, 2014).Traditional agriculture which is based on rigorous soil tilling has resulted in soil erosion problems, surface and underground water pollution, and more water consumption. Land degradation, wildlife and biodiversity reduction, low energy efficiency and contribution to global warming problems (Boatmann *et al.*, 1999) are the problems associated with conventional agricultural practices. Conservation agriculture (CA) offers several advantages in terms of crop residue management and cover crops, in order to offer a permanent soil cover and a natural increase of organic matter content in surface horizons.

CA is a suite of land and cropmanagement practices to improve productivity,profitability, and sustainability. Conservation agriculture builds a more integrated and holistic approach onto soil management and soil productivity, fertilizer, organic matter, water management and field preparation.

Conservation Agriculture – The concept

Conservation agriculture is a management system that maintains a soil cover through surface retention of crop residues with no till/ zero and reduced tillage. CA is a concept for resource saving agricultural crop production which is based on enhancing the natural and biological processes above and below the ground. CA is based on optimizing yields and profits, to achieve a balance of agricultural, economic and environmental benefits. It advocates that the combined social and economic benefits gained from combining production and protecting the environment, including reduced input and labor costs, are greater than those from production alone. It aims to conserve, improve and make more efficient use of natural resources through integrated management of available soil, water and biological resources combined with external inputs. It can also be referred to as resource efficient or resource effective agriculture.

Conservation Agriculture (CA) is an approach to managing agro-ecosystems for improved and sustained productivity, increased profits and food security while preserving and enhancing the resource base and the environment. CA is characterized by three linked principles, namely:

- Continuous minimum mechanical soil disturbance.
- Permanent organic soil cover.
- Diversification of crop species grown in sequences and/or associations.

Efforts are continuously being made to develop, refine and disseminate conservation-based agricultural technologies for nearly two decades. Conservation agriculture based technologies have made a significant progress in the recent past. Particularly, tremendous efforts have been made on no-till in wheat under a rice-wheat rotation in the Indo-Gangetic plains. The technologies of CA provide opportunities to reduce the cost of production, save water and nutrients, increase yields, increase crop diversification, improve efficient use of resources, and benefit the environment.

India is a large country with very diverse and multiple cropping systems. For sustainable intensification of rice-wheat cropping system and incorporation of best management practices such as timely crop establishment, zero-tillage (ZT) was introduced and was also accepted by farmers at scale. That required a mind-set shift which did not come through research trials but through farmer participatory on-farm trials. Zero-tillage was introduced as a strategy to reduce the tillage cost so that farmers can invest in purchasing new herbicides and for reducing turnaround time between rice harvest and wheat planting. The adoption of this technology has now offered new opportunities in the field of cropping system intensification especially in eastern Indo-Gangetic Plains of India. On-farm participatory research approach is effective in changing the mind set of research organization and farmers. Zero-tillage helps in early wheat sowing, adoption of long duration wheat, using the available residual moisture and optimization of cropping systems. Zero tillage research which was at the dead end in 1996 in NW India and in EIGP early wheat sowing in the first two weeks of November was at the dead end till 2010-11. The farmer's participatory approach changed the perceptions in both cases.

Zero Tillage (ZT)

The ZT now is best-targeted technology for the eastern Indo-Gangetic Plains (EIGP) where cropping system optimization demand advancement in the wheat establishment to beat the terminal heat stress and for greater system productivity. CSISA in the EIGP is focusing on creating a critical mass of service providers so that small and medium farmers who face cash-flow risks and who cannot buy machines can also have the access of ZT technology through custom-services provided by service providers (SPs) who own such machines. Moreover, farmers in EIGP have also started facing the problem of Phalaris minor due to late wheat sowing when temperature favorsP. minor emergence compared to early planting. Zero tillage will be guard against such eventuality as the early wheat sowing mediated through zero tillage will discourage this weed to dominate the wheat crop.

ZT in wheat reduces the emergence rate of phalaris minor compared to conventional tillage (CT) (Franke *et al.*, 2007). The lower emergence of P. minor under ZT may be attributed to (1) higher soil strength in ZT because of crust development in the absence of tillage after rice harvest, which can mechanically impede seedling emergence (Chhokar *et al.*, 2007), and (2) higher weed seed predation under ZT (Kumar *et al.*, 2013). Other possible factors could be (1) less soil temperature fluctuation because ZT helps in moderating soil temperature (Gathala *et al.*, 2011) or (2) lower levels of light stimuli, N mineralization, or gas exchange, all of which are known to stimulate germination of many weed species following tillage (Franke *et al.*, 2007). In an impact assessment study conducted recently in the eastern IGP in Bihar by CSISA project, the benefits of ZT include 9.3% higher yield (0.24 t/ha) and US$ 74 more net revenue per ha compared to CT (Alwin *et al.* unpublished). Other benefits of ZT compared to CT include additional tillage cost reduction, seed rate maintenance thus saving seeds and irrigation water saving (13-33%) and improvement in soil health.

Direct Seeded Rice (DSR)

Puddled transplanted rice (PTR)

Puddled transplanted rice is the most common method of rice establishment. Puddling was done by farmers over the centuries for very specific reasons, the most important being to help control weeds; farmers found that keeping soils anaerobic and flooded reduced weed problems; but also hand weeding was easier with these softened soils. The puddling essentially reduced water percolation and infiltration and ponded the water on the surface. Based on the work done by Reichardt *et al.* (1997) and Reichardt *et al.* (2001) concluded that soil microbial biomass plays a significant role as a passive nutrient pool and suggests that its reduction found in puddled soils in the second half of the cropping season could be a mechanism that contributes to declining productivity in continuous rice cropping systems. Now labour availability has also become a serious problem and to ensure timely transplanting Direct Seeding of Rice and mechanical transplanting are another viable option even under non-puddled conditions (MTNPR) with minimum disturbance of soil. The weed management in MTNPR is same as in the PTR but yields are generally higher due to transplanting of young seedling.

Direct Seeded Rice (DSR)

Direct Seeded Rice if practiced with correct management practices where assured irrigation and timely spray of herbicides in the correct combination and right dosage is a requirement. In spite of best efforts during last 20 years, we have seen weakest growth in area under direct seeded rice (DSR) throughout except some successes in basmati rice areas of Haryana and also in low productivity areas of Punjab. Two factors were important: one competitive varieties and the other water with assured irrigation. Both these factors are not prevalent in the Eastern Indo-Gangetic Plains (EIGP).

The yield growth and net profit growth is affected by weed competition in DSR. Herbicides when applied alone do not provide full season weed control. This has put DSR farmers under pressure to use alternate methods like hand weeding which is becoming costly and scarce. Some herbicides for nut sedge (*Cyprus rotundus*) dominated complex weed flora have been tested. Herbicides like pyrazosulfuron applied alone at 25 or 30g/ha or halosufuron applied 30-35 DAS at 60g/ha or bispyribac applied 20-30 DAS at 30g/ha can control nut sedge. None of these herbicides will control the complex weed flora dominated by nut sedge. The tank mixture of 20g pyrazosulfuron + 20g /ha bispyribac or halosulfuron is very effective against nut sedge dominated complex weed flora of DSR. Some weeds like *Leptochloa chinensis, Eragrostis tenuifolia* and *Dactyloctenium aegyptium* are not controlled by bi-spyribac. For these weeds azimsulfuron or cyhalofop-butyl will have to be tank mixed with bispyribac or fenoxaprop + safener (Rice Star) needs to be applied. Oxadiargyl at 90 g/ha applied as tank mixture with bispyibac provide long season weed control of annual grasses which emerge after the treatment of bispyribac. Pre-emergence spray of pendimethalin, pretilachlor +safener (Sofit) or oxadiargyl depending upon soil moisture followed by post- emergence spray of bispyribac also controls major weed flora in DSR or MTNPR including *Leptochloa chinensis, Eragrostis tenuifolia* and *Dactyloctenium aegyptium*.

Development of effective weed management strategies for DSR has played an important role in the expansion of area under DSR. Diverse and complex weed flora and prolonged weed emergence pattern contribute to the complexity of weed management in DSR. In North West IGP, integrated weed management strategies based on herbicides and manual weeding has been successful in DSR. Based on on-farm and on-station trials, bispyribac-sodium 25 g/ha sprayed at 15-25 days after sowing (DAS) was extremely effective against *Echinochloa* species and some broad-leaf weed (BLW) and sedges in DSR and transplanted rice. Tank-mix of azimsulfuron 20 g/ha or pyrazosulfuron 25 g/ha with bispyribac-sodium 25 g/ha has also provided excellent control of complex weed flora including BLW and sedges including purple nut sedge. Azimsulfuron alone also provided effective control of most BLW and sedges. Halosulfuron alone at 60 g/ha was found excellent on sedges including *C. rotundus*. Many researchers have reported that pendimethalin (pre-emergence) followed by post-emergence application of bispyribac or azimsulfuron or bispyribac-sodium + azimsulfuron 15-20 DAS yielded similar to weed-free conditions.

Weedy Rice in Direct Seeded Rice

Weedy rice and its development into an important problem is associated with low land rice ecology of Eastern-UP and adjoining parts of Bihar. Stale bed technique to exhaust the existing seed bank and use of hybrid seed to solve seed contamination problem should be an effective strategy. This will help facilitating the adoption of DSR especially under conventional tillage (CT). Like any other technology such practicalities may get in the way forward. If we look at the whole system, the use of pre-seeding herbicides can be an efficient tool which can lead to flexibility in respect of using or not using post emergence herbicides. This also makes it possible to boost the early crop canopy cover. Frequent use of herbicides in zero tillage system seems to be only an ideological obstacle in the process of tillage reforms. Given the fact that we have a very high weed population of complex weed flora in EIGP, excellent crop cover mediated by hybrid rice and sowing under CT with possibility of creating soil mulch will help simplifying the weed flora by creating the competition in favor of crop. The utility of soil mulch is more under rain-fed ecologies especially when rains are not expected 2-3 weeks after seeding rice. The success of soil mulch could be in part due to less emergence of weeds from the dry surface soil and also conservation of soil moisture due to disruption of evaporation process by dry surface.

Constraints for adoption of CA

CA is now, considered a route to sustainable agriculture. Spread of conservation agriculture, therefore, will call for scientific research linked with development efforts. The following important constraints which impede broad scale adoption of CA have been identified by Bhan and Behera (2014).

- Lack of appropriate seeders especially for small and medium scale farmers: Although significant efforts have been made in developing and promoting machinery for seeding wheat in no till systems, successful adoption will call for accelerated effort in developing, standardizing and promoting quality machinery aimed at a range of crop and cropping sequences. These would include the development of permanent bed and furrow planting systems and harvest operations to manage crop residues.

- The wide spread use of crop residues for livestock feed and fuel: Specially under rainfed situations, farmers face a scarcity of crop residues due to less biomass production of different crops. There is competition between CA practice and livestock feeding for crop residue. This is a major constraint for promotion of CA under rainfed situations.
- Burning of crop residues: For timely sowing of the next crop and without machinery for sowing under CA systems, farmers prefer to sow the crop in time by burning the residue. This has become a common feature in the rice-wheat system in north India. This creates environmental problems for the region.
- Lack of knowledge about the potential of CA to agriculture leaders, extension agents and farmers: This implies that the whole range of practices in conservation agriculture, including planting and harvesting, water and nutrient management, diseases and pest control etc. need to be evolved, evaluated and matched in the context of new systems.

Skilled and scientific manpower: Managing conservation agriculture systems, will call for enhanced capacity of scientists to address problems from a systems perspective and to be able to work in close partnerships with farmers and other stakeholders. Strengthened knowledge and information sharing mechanisms are needed.

❑❑❑

13

Crop Improvement for Water and Heat Stress Tolerance

It is estimated that by 2025 the water use in the agriculture sector is expected to increase by 1.3 times (Shiklomanov, 1999). To meet the water requirements of the burgeoning population and for the industrial consumption, water has to be diverted to other uses leaving less water available for agriculture. Also, emission of greenhouse gases such as carbon dioxide, methane and nitrous oxide emission from agricultural fields are major sources contributing to the global increase in temperature. These changes are bringing in extreme events in climatic conditions such as excess rain in a short span, excess increase in temperature, sudden drop in temperature, cloudy conditions reducing available sunshine, drought and flood. Although it is predicted that slight increase in carbon dioxide is favorable for yield and slight increase in temperature would increase crop yields in temperate areas, crop plants have started experiencing spans of contrasting stresses. Thus, developing plants with multiple stresses has become a great challenge. The following section deals with two majors stress *viz.*, water deficit and heat and challenges in developing tolerant varieties. Under deficit water supply, the survival and yield prospects of the crops are severely affected. There is need to explore the possibilities for saving available water for meeting the requirements of critical growth stages of the crop production.

Crop Response to limited water supply

The adoption of plants to water deficit is a very complex process involving morphological, physiological, biochemical, and molecular alterations. The drought tolerance can be observed with the traits related to the osmotic adjustment, cell membrane stability, epicuticular wax on the plant parts, ability for partitioning the stem reserves to the economic parts of the yield, manipulation and stability of flowering processes etc. In the short-term soil drought, yield reduction could be related to reduced stomatal conductance and concomitantly lower photosynthetic rates but in the long term, a smaller leaf area (due to decreased leaf size and higher rates of leaf senescence), and an altered assimilate partitioning between plant structures and organs could be more directly responsible for reduced yield.

Stomatal Conductance: Reductions in stomatal conductance happen due to internal water deficits as drought progresses and, as a consequence, yield reduction happens. Another problem associated with reduced stomatal conductance (and thus reduced transpiration rate) is increased leaf temperature, which has possible consequences on higher maintenance respiration and photorespiration rates, which further contributes to decrease in crop productivity. Variation in inherent water use efficiency exhibited by some of the genotypes is due to the reduction in stomatal conductance (g) rather than increased photosynthetic capacity at a given stomatal conductance (Meinzer *et al.*, 1990). In case of Soybean, plant stem elongation, stomatal conductance, relative water content, water potential, osmotic potential and turgor pressure of stress plants declined in all genotypes when studied under water stress (Hossain *et al.*, 2015). In crops like rice, high dry matter production of the drought resistant cultivars is not due to high water use efficiency, but due to their high ability to maintain transpiration, which is supported by deep root systems. (Pinheiro *et al.*, 2005, Kobata *et al.*, 1996).

Root Development: In addition to factors leading to a lower transpiration rate, the adaptation to water deficit involves a greater capacity for water absorption by the plant. This process can be accomplished with genotypes that have more robust root systems (deep and branched), which can better exploit deeper soil water reserves. Indeed, a hallmark of plants subjected to water stress is an increase in the biomass allocated to the root system. Plants respond to water deficit using mechanisms of avoidance by improving root traits (Price *et al.*, 2002). Though osmolyte accumulation (OA) is generally considered as a mechanism for increasing crop yields under drought conditions, the field studies examining the association between OA and crop yield had not shown any benefit (Serraj and Sinclair 2002). However, mechanism identified for beneficial yield responses to OA is the maintenance of root development in order to reach water that may be available deeper in the soil profile. Yield is a function of water used, water-use efficiency and harvest index and during the water stress, the crops have access to a substantial amount of water, but they have to exploit with the root system to mine to residual water to improve the yields (Passioura 1983).

Leaf parameters: Under moderate water deficit conditions, decreases in leaf area are not necessarily accompanied by decreases in photosynthetic rates on a leaf area basis, but if drought progresses further, strong decreases in photosynthetic rates per unit leaf area are often observed. All the physiological traits such as leaf area index, leaf area ratio, specific leaf area and the leaf weight ratio decrease with decreasing water availability (Farshbaf-Jafari *et al.*, 2014). Other important traits like plant height and root length, number of pods produced per plant are severely affected under soil moisture stress at the flowering stage (Dahanayake *et al.*, 2014).

Crop Yield: Under water stress conditions, the yield (Y) of a crop can be described by the equation: Y = E x WUE x HI, where E = transpired water, WUE = water-use efficiency and HI = harvest index (i.e., relationship between the biomass of the organ of commercial interest and the total plant biomass). Large increases in productivity through breeding programs occurred by increasing the harvest index. For most annual species grown; however, HI is close to its maximum value, and additional increases in crop yields must necessarily be accomplished through increased biomass accumulation. Therefore, the challenge is to increase the capacity of the plant to produce a larger amount of dry matter per unit area.

Crop Response to heat stress

Excessive temperature and high irradiance, are common under drought conditions. These factors explain why drought is considered to be a multidimensional stressor. With a reduction in photosynthesis, the use of incident solar radiation decreases, which can damage plant metabolism through processes such as lipid peroxidation and protein and nucleic acid oxidation. These oxidative damages can be observed in leaves as chlorotic areas, which in many cases progress to necrosis, leading to leaf abscission. Genotypes with higher antioxidant activity, both enzymatic (increased activity of enzymes, such as superoxide dismutase, catalase, ascorbate peroxidase, and glutathione reductase) and non-enzymatic (e.g., glutathione and ascorbate) tend to be more tolerant to drought, snowing less oxidative damage compared with genotypes with less robust antioxidant systems. This high antioxidant capacity confers increased cellular protection, especially for the photosynthetic machinery, and maintenance of leaf area, which consequently results in increased production, even under severe water deficits.

In breeding programs, selection for higher WUE for water-limited conditions would lead to increased yield and also drought resistance. Till the time biochemistry of photosynthesis is improved genetically, transpiration efficiency and WUE would lead to a reduction in transpiration and crop water-use which are very crucial for plant production. Therefore, breeding for efficient soil moisture capture traits should be the important target for yield improvement under drought stress.

Breeding for water deficit stress

Breeding programs for drought conditions must combine increased biomass production with lower water consumption or increased WUE. Reducing the stomatal conductance, increasing the photosynthetic capacity or even combining both factors may lead to a higher WUE. A reduction in stomatal conductance is not favorable in breeding programs aimed at increasing productivity because it entails less CO_2 influx and lower photosynthetic rates and, therefore, less biomass accumulation. Thus, the great challenge is to increase the photosynthetic capacity under conditions of low stomatal conductance. Several strategies have been proposed to increase the photosynthetic capacity of cultivars. One of these strategies involves CO_2 concentrator mechanisms, such as the one found in species that utilize C_4 metabolism. Another strategy would be to increase the CO_2 specificity of the enzyme ribulose 1, 5- bisphosphate carboxylase/oxygenase (Rubisco), thus reducing losses related to photorespiration in C_3 plants.

The increase in the mesophyll conductance is directly associated with increases in photosynthetic rates without the need for increased stomatal conductance. Finally, another strategy involves increasing the specific leaf mass because such an increase represents a greater amount of the photosynthetic apparatus per leaf unit area.

At a cellular level, osmotic adjustment has been considered as the most striking response of plants subjected to water deficit, which is characterized by a net solute increase in cells. These solutes, named compatible osmolytes, are organic compounds (e.g., proline sugar alcohols, sorbitol, mannitol) that do not interfere with cell metabolism. The osmotic adjustment has two main functions: (i) maintaining cell turgor in the presence of negative water potential, which allows the process of cell elongation to be maintained in addition to allowing relatively larger stomatal aperture and (ii) providing a greater capacity for water uptake by the roots, which would also contribute in the maintenance of higher stomatal conductance. Concomitantly, osmotic adjustment leads to the maximization of the photosynthetic process, resulting in relatively greater biomass accumulation. Thus, genotypes with greater potential for osmotic adjustment tend to be more tolerant and, therefore, more productive under drought conditions (Pattanayak, 2014).

Drought or water deficit is defined as a condition of moisture deficit sufficient to have an adverse effect on vegetation. Premature senescence, abscission of flowers, delay in flowering & grain formation, spikelet sterility etc. are some of the indicators of drought in crop plants. Drought tolerance is defined as the ability of a genotype to be more productive with a given amount of soil moisture than another genotype. Drought was classified as Terminal, Unpredictable and Predictable & terminal based on the stage of occurrence (Jordan and Miller, 1980).

Water deficits develop due to loss of water through stomata. Water lost is replaced by water drawn from soil through root, stem and leaf via xylem, which was explained by the phenomenon Soil-plant-atmosphere continuum (Passioura, 1982). The activity of plants with a water deficit depends upon climatic conditions, which affect the quantity of water available and how the water is used. The responses to limited water availability are diverse with respect to plant species involved and/or the severity and duration of the water deficit. There are three strategies by which plants can grow and develop properly with water deficit conditions:

Drought escape (Turner, 1986): It is the ability of the plant to complete its life cycle before soil water is depleted. Phenology of the crop matches with the soil moisture availability. Short duration, earliness, dormant seeds etc are some of the examples. It is usually observed under terminal drought condition.

Drought avoidance (Levitt, 1980): It is the ability to maintain high tissue water potential despite shortage of soil moisture. Optimal use of water is obtained for the production of dry matter. Drought can be avoided by two means either by reducing transpiration or increasing absorption of water:

- Water Savers: reduced transpiration by closure of stomata (Jones, 1998), leaf drop (Turner, 1986) and reduced surface by smaller thicker leaves
- Water Spenders: increased absorption of water by improving capacity of root system (Turner, 1986 and Subbarao *et al.*, 1995).
- Stress is excluded by means of morphological, physiological and biochemical barriers such as:
- Morphological: reduced leaf area, leaf rolling, cuticular wax, deep rooting system, awn, stability in yield
- Physiological: reduced transpiration, high water use effeciency, stomatal closure, osmotic adjustment
- Biochemical: accumulation of proline, polyamine, trehalose, increase nitrate reductase activity and increased storage of carbohydrate.

Drought tolerance: It is the ability to withstand water deficit with low tissue water potential. This is obtained by maintenance of turgor through osmotic adjustment (induces solute accumulation), increases elasticity in the cell and decrease in cell size. It involves cellular tolerance mechanism. Drought recovery which involves increased vegetative growth vigour is also a tolerant mechanism.

Each of these tolerances is associated with costs and benefits, which vary according to the species, the environment, the technological resources of the farmer, and the goals of breeding.

Some of the successful approaches in drought tolerance:

- P5C5 gene codes for rate limiting enzyme in proline biosynthesis pathway from cowpea and overexpressed in tobacco resulted in P5C5 transgenics highly tolerant to stress (Kavikishore *et al.*, 1995).
- SHINE/WIN 1 AP2 ERF transcription factors (Asaph *et al.*, 2004), WXP 1 (Zhang *et al.*, 2005) have been successfully found to increase wax synthesis which improved drought tolerance.

- Transcription factors At DREB 1A in groundnut, rice, potato and pigeon pea (Shinozaki *et al.*, 2002), Os DREB 1A in rice (Shinozaki & Shinozaki, 2005), DREB 1A in soybean and vegetables, HsfA1 in tomato (Nover *et al.*, 1983), SNAC in rice (Xiong *et al.*, 2006), Plant nuclear factor NF-Y in maize (Nelson *et al.*, 2007), Bacterial RNA chaperon csp-A in maize were validated under field conditions.
- Improvement of water use efficiency in rice by expression of *HARDY*, an *Arabidopsis* drought and salt tolerance gene (Aarti *et al.*, 2007).
- Consistent drought grain yield QTLs (qDTY1.1, 2.2, 3.1, 3.2, 4.1, 6.1, 12.1)) identified through bulk segregant analysis in N22 and Dhagaddeshiof rice (Vikram *et al.*, 2012).
- IRRI drought tolerant breeding programme of rice involved transfer of identified QTLs (DTY12.1, 3.1) into mega varieties IR64, MTU1010, Swarna, SambhaMahsuri, Anjali and Vandana by AB QTL approach and 4 QTLs (DTY1.1, 2.1, 3.1, 8.1) in Swarna and Swarna Sub1by gene pyramiding approaches. Table 1 describes IRRIs's successful drought breeding programme in South, South-East Asia and Africa.

Table 13.1 High-yielding drought-tolerant varieties developed from IRRI's and released in different countries of South and Southeast Asia and Africa

Name	Designation	Days to maturity	Plant height (cm)	Country, release year, situation
Sahod Ulan 1	IR74371-54-1-1	110	104	Philippines 2009, RL, UP
Hardinath 1	IR80411-B-49-1-1	115	100	Nepal 2009, RL
Sahbhagidhan	IR74371-70-1-1	110	104	India 2010, RL, UP
BRRI dhan56	IR74371-70-1-1	110	108	Bangladesh 2011, RL
Sookhadhan 3	IR74371-70-1-1	110	108	Nepal 2011, RL
Sookhadhan 1	IR74371-46-1-1	110	101	Nepal 2011, RL
Sookhadhan 2	IR74371-54-1-1	110	104	Nepal 2011, RL
Katihan 1	IR79913-B-176-B-4	105	90	Philippines 2011, UP
Sahod Ulan 3	IR81412-B-B-82-1	120	107	Philippines 2011, RL
Sahod Ulan 5	IR81023-B-116-1-2	115	130	Philippines 2011, RL
Sahod Ulan 6	IR72667-16-1-B-B-3	115	100	Philippines 2011, RL
Sahod Ulan 8	IR74963-262-5-1-3-3	125	100	Philippines 2011, RL
Inpago LIPI Go 1	IR79971-B-191-B-B	110	115	Indonesia 2011, UP

Inpago LIPI Go 2	IR79971-B-227-B-B	113	114	Indonesia 2011, UP
Sahod Ulan 12	IR81047-B-106-2-4	105	119	Philippines 2013, RL, DS
M'ZIVA	IR77080-B-B-34-3	120	130	Mozambique 2013, RL
UPIA3	IR74371-54-1-1	110	104	Nigeria 2013, RL

Vegetable breeding for drought tolerance is still at infancy followed by traditional methods of identification of tolerant genotypes and conventional breeding methods to transfer the tolerance. Table 2 compiles drought tolerant species in vegetables (Kumar *et al.*, 2012).

Table 13.2 Drought tolerant species & genotypes of vegetables

Vegetable	Drought tolerant species/genotypes	References
Tomato	*S. habrochaites* (EC- 520061), *S. pennelli* (IIHR 14-1, IIHR 146-2, IIHR 383, IIHR 553 , IIHR 555, K-14, EC-130042, EC-104395, Sel-28), *S. pimpinellifolium* (PI-205009, EC- 65992, Pan American), *S. Esculentumvar. cerasiforme, S. hirsutum, S. cheesmanii, S. chilense, S. habrochaites, S. Sitiens*	Rai *et al.* (2011)
	Arka Vikas, RF- 4A	Singh (2010)
	L. pennellii (LA0716), L. chilense (LA1958, LA1959, LA1972), S. Sitiens (LA1974, LA2876, LA2877, LA2878, LA2885), *S. Pimpinellifolium* (LA 1579)	Razdan & Mattoo (2007)
Brinjal	*S. microcarpon, S. gilo S. macrosperma, S. integrifolium, Bundelkhand* Deshi *S. sodomaeum (syn. S. linneanum)*	Toppino *et al.* (2009) Rai *et al.* (2011)
	SM-1, SM- 19, SM- 30, Violette Round, Supreme	Kumar & Singh (2006)
Chilli	*C. chinense, C. baccatum var. pendulum, C. eximium* ArkaLohit, IIHR - Sel.-132	Singh (2010)
Potato	*S. acaule, S. demissum and S. stenotonum, Alpha, Bintje S.ajanhuiri, S.curtilobum, S.xjuzepczukii* KufriSheetman	Arvin & Donnelly (2008) Ross (1986)
	Solanum chacoense, Kufri Sindhuri	Pandey *et al.* (2007)
Okra	*A. caillei, A. rugosus, A. tuberosus*	Charrler (1984)
Onion	*Allium fistulosum, A. munzii, Arka Kalyan,* MST 42, MST 46	Singh (2010)
French bean	*P. acutifolius*	Kavar *et al.* (2011)
Sweet potato	VLS6, IGSP 10, IGSP 14, Sree Bhadra	Singh (2010)
Cassava	CE-54, CI-260, TP White, Narukku-3, Ci-4	Singh (2010)
Watermelon	*Citrullus colocynthis (L.) Schrad*	Dane *et al.* (2007)

Cucumber	INGR-98018 (AHC-13)	Rai *et al.* (2008)
Winter squash	*Cucurbita maxima*	Chigumira & Grubben (2004)
Cucumis Spp.	*Cucumis melo var. momordica* VRSM- 58, INGR-98015 (AHS-10), INGR-98016 (AHS-82), CU 159, CU 196 *Cucumis pubescens*, INGR-98013 (AHK-119) *Cucumis melo var. callosus*, AHK- 200, SKY/DR/RS-101 *Cucumis melo* var. *chat, Arya Cucumismelo*, SC- 15	Rai *et al.* (2008), Kusvuran (2012), Pandey *et al.* (2011)

The progress in the last 100 years, in relation to obtaining drought tolerant cultivars, has been very weak in most breeding programs, regardless of the species considered. Many experiments have been conducted in greenhouses or growth chambers with plants in small pots, a condition that causes the plants to be exposed to drastic and swiftly imposed drought which do not entirely reflect the field situation, and therefore, the comparative results should be treated with caution. Consequently, much of the drought tolerance information generated by physiologists has limited utility for the crop breeder because such information cannot be converted to adequate indices to predict the behavior of genotypes with varying levels of tolerance to water stress.

The dissociation between basic information and the requirements of applied (or applicable) information is an additional bottleneck for advancements leading to the acquisition of more drought-tolerant materials. There is great need to conduct multidisciplinary research involving molecular, physiological, and classic genetic improvement to enhance crop performance in drought prone regions. Recent advances in the transcriptome, proteome, and metabolome have allowed an integrated view of the expression of thousands of genes and their products involved in drought tolerance. In the long term, these improvements will provide a better understanding of the polygenic nature of drought tolerance. The adequate selection and parameters for the identification of genetic material with increased water deficit tolerance has resulted in satisfactory increases in the yield of some crop. Improved WUE, more robust root systems, osmotic adjustments, and increased activity of the antioxidant system are some of the features that should be on the breeding programs aiming to increase production under water deficit condition.

Breeding for Heat Stress

The threshold temperature (upper and/or lower), which varies both inter and intraspecifically, refers to the daily average temperature that causes a reduction in growth. The temperature increase above the upper limit for a period of time is sufficient to cause irreversible damage to plant growth and development is defined as heat stress, which is a complex function of the intensity, duration, and rate of temperature increase. It leads to death of cells, tissues, organs and whole plants. Sun scald in tomatoes and grapes, bark burn in trees are some of the examples of heat stress. Plants are subjected to heat stress under the following conditions: (i) the air temperature is high, and the plants receive energy by the transfer of sensible heat; (ii) solar radiation incident on the ground raises the temperature above the air temperature; and (iii) solar radiation-induced heating and the inability to dissipate heat can rapidly warm the leaves (up to 15°C above the air temperature or more), particularly in leaves that have a low transpiration rate and are frequently subjected to high temperatures.

Wilting, leaf burn, leaf folding or abscission are some of the indicators of heat induced damage. Increased transpiration rate, higher leaf reflectance or angle or orientation to the sun are examples of heat avoidance in crop plants. Greater water uptake, thicker leaves, higher leaf sap electrical conductivity and chlorophyll content and lower stomatal number are related to heat tolerance which operate at cellular level.

Heat stress can affect crop yield at any time from sowing to grain maturity, but it is the time around flowering, when the number of grains per land area is established, and during the grain-filling stage, when the average grain weight is determined, that high temperatures have the most impact on the final harvestable crop, as found in cereals.

Effect of high temperature on flower initiation: Temperatures > 30^{0}C during floret formation cause complete sterility in wheat (Saini and Aspinall, 1982). In wheat, number of kernels/unit area decreases @ 4% for each degree increase in mean temperature during anthesis (Fischer, 1985). In rice, floral abnormalities induced by heat stress leads to spikelet sterility (Takeoka *et al.,* 1991).

Effect of heat stress on fertilization: In wheat, high temperature stress (>30 ^{0}C) from early meiosis to pollen maturity has a damaging effect on the viability of pollen grains resulting in failure of fertilization and thus reduction in seed set (Saini and Aspinall, 1982). Increased temperature over mid anthesis period decreased the grain number/ ear at maturity in spring wheat (Ferris *et al.*, 1998). Sterility is caused by poor anther dehiscence and low pollen production and hence low number of germinating pollen grains on the stigma (Prasad *et al.*, 2006). In maize, reduction in seed set occurs at temperatures higher than 38^0c mainly because of a reduction in pollen germination ability and pollen tube elongation (Stone, 2001).

Grain filling under heat and drought: Large yield losses in cereals because of reduction in starch accumulation. In wheat, drought (20 days at endosperm cell division) resulted in 30-40% lower endosperm cell number, No. of small starch granules per cells was also reduced by 45% (Nicolas *et al.*, 1985).

Thus, understanding how environmental factors signal phenological processes such as flowering will be extremely relevant for future food production, since a large part of food comes as grains/ seeds. In this context, any change in flowering time could affect not only seed production but also food composition. The reduction in grain weight in response to heat stress during the initial stage of filling can be partially attributed to the small number of cells in the endosperm, while the subsequent stress reduces starch synthesis due to the reduced availability of assimilates for the grain or direct effects on the synthesis processes in the grains. Photosynthesis is limited at temperatures above 35 °C due to demand, which normally occurs at high temperatures, contributes to the reduction of photosynthesis. In addition, stomatal closure due to high evaporative air demand, which normally occurs at high temperatures, contributes to the reduction of photosynthesis.

Most plants have a considerable capacity to adjust photosynthetic performance to their growth temperatures. Part of that capacity can be assigned to the properties of Rubisco, which is very thermostable. Notably, the rate of carboxylation may increase at 40 °C or above but at a lower speed than the oxygenation rate; consequently, higher photorespiration rates are frequent at high temperatures, which ultimately reduce carbon gain. Therefore, Rubisco that has more carboxylation capacity would be advantageous in situations of excessive heat. Regardless, at moderately high temperatures, the low activation slate of Rubisco would be the main cause of the decreased photosynthetic rate due to the suppression of Rubiscoactivase activity. The accumulation of compatible osmolytes is an important adaptive mechanism to heat stress. Under heat, different plant species can accumulate a variety of osmolytes, such as sugars, sugar alcohols, proline, and betaines, as a way to increase tolerance to heat stress, such as in water and salt stresses. Due to the pivotal role of osmolytes in response to environmental stresses, the selection of materials with a greater potential for the accumulation of these osmolytes, whether through traditional genetic breeding, marker assisted selection or genetic engineering, is of special interest.

In summary, the measurement of membrane thermal stability is a good indicator of high temperature tolerance. In this context, the electrolyte leakage method, which is used to check the loss of cell compartmentalization, seems very sensitive for the quantification of temperature tolerance, as demonstrated in rice. To increase heat tolerance, crop breeding has been conducted by biotechnological methods such as the control of membrane composition or creation of cultivars with the constitutive expression or overexpression of heat shock proteins. Studies have indicated that plant tolerance to heat is a multigenic characteristic. Despite the genetic complexity and difficulties heat tolerant strains and hybrid cultivars have been developed in crops such as rice and maize. Changing plant architecture to increase planting density and thereby increasing canopy density can reduce thermal stress with minimal effect on water consumption (Pattanayak, 2014).

Transitory or constantly high temperatures cause an array of morphoanatomical, physiological, and biochemical changes in plants, which affect plant growth and development and may lead to a drastic reduction in economic yield. The adverse effects of heat stress can be mitigated by developing crop plants with improved thermotolerance using variousgenetic approaches (Wahid *et al.*, 2007). Lucas *et al.*, 2013 discovered five QTLs for breeding heat tolerant cowpea. N22 and Dular are excellent sources of genes for heattolerance in rice (Manigbas *et al.*, 2014).

Field screening for drought / heat tolerance is complicated by the fact that both are highly influenced by genotype x environment interaction. Choosing uniform sites for evaluation, more replications and larger set of germplasm are some of the strategies followed by the breeders. In addition, the selection environment should be a representative of the target environment and there should be a good genetic correlation between the selection environment and target environment. A sequential path analysis of the component characters may help to identify the proper predictor variables.

Stress tolerance technology could help farmers get the most from their inputs and management practices and protect their investments in times of water shortages and high temperatures.

□□□

14

Pests, Diseases and Weed Management

Water productivity implies the net benefits (yield, economic returns) achieved per unit of water used by the crop or applied to the field. The numerator is net benefit in which crop yield is the major factor. Water productivity can be increased either by increasing the numerator (yield) or reducing the denominator, i.e. water use. The focus is to get more yield and income per unit of water. Yield is central to any crop production activity. Besides better soil and water management, the yields are significantly affected by insect pests. Insect pests, diseases and weeds inflict enormous losses to the potential production in rainfed ecosystem. Anecdotal evidences also indicate rise in the losses, despite increasing use of chemical pesticides. It is estimated that herbivores pests eat about 20% of the crop grown for human consumption. Development of resistance, secondary pest outbreaks and emergence of new pest problems is causing huge monetary loss. The annual losses due to insect pests and diseases vary up to 38 % (insect pests and diseases 26%, weeds 10% and birds 1-2 %). Reduction in yield leads to reduced water productivity indicating that proper crop protection measures has to play a key role in improving water productivity. Also, expenses on application of pesticides, herbicides and fungicides reduces the net farm income which further adds to reduction in economic water productivity.

Management of pests, diseases and weeds is critical to improve the water productivity of agriculture.

Integrated Pest Management

With the adoption of intensive agricultural practices, there has been a major change in the status of several pests of field and horticultural crops in rainfed ecosystem in the recent past. Many minor pests have now assumed the status of major pests. Even after introduction of resistant cultivars, evolution of new biotypes has become a regular phenomenon. Mites, thrips and whiteflies, which were insignificant in rainfed ecologies, have become a threat. However these negative externalities cannot be eliminated altogether. Their intensity can be minimized through development, dissemination and promotion of alternative technologies such as biopesticides and bioagents as well as good agronomic practices rather relying solely on chemical pesticides.

Pest management strategies for rainfed ecosystem should consider the fact that multiplication of pests is very fast under rainfed conditions. Poor and resource limited farmers cannot afford pesticides and fungicides at right time and they have to resort to alternative measures that are indigenous, economical and environment-friendly. Pest management stratifies for some important crops of the region are provided here.

Rainfed Paddy

Paddy is the main crop in rainfed ecosystem which suffers by many pests. Owing to cultivation of semi dwarf varieties and intensive cultivation, there has been a major change in the status of several paddy pests in the recent past (Pathak, 1986). Major pests of rainfed ecosystem are Yellow stem borer (YSB), Brown Plant Hopper, Green Leaf Hopper, gall midge, Hispa, Leaf folder and Gundhi bug (Litsinger, 1979; Reissig *et al.*, 1986). Mites and nematodes, which were insignificant in rainfed ecologies, have become a threat in recent years. Strategies of pest management in rainfed paddy ecosystem to keep the pest population below threshold level are elaborated below:

Yellow Stem Borer: *Scirpophaga incertulas* (Wlk.)

The caterpillar of borer is yellowish white with orange in coloured head. The adult moth can be easily identified which body is slender in shape, straw in coloured with a pointed head. The female moth has a distinct black spot at the centre of fore wings (Bora *et al.*, 1994). Caterpillars which feeding inside the paddy stem causing "dead heart" or drying of the central whorl of leaf during the tillering stage. If the attack occurs during panicle formation or heading stage then "white ear heads" are formed and the whole panicle becomes chaffy. The incidence can be reduced by adopting following management strategies:

- Summer deep ploughing of field to expose the resting stages of insects to heat of the sun and natural enemies.
- Early and synchronous transplanting planting reduce the incidence.
- Balance fertilizer application avoiding excess use of nitrogen.
- Field should be free from weeds to check carryover of pests.
- Avoid of mono-cropping
- Regular surveillance is required to paddy fields for pests and parasitoids.
- Mechanical killing of adults through light trapping and destruction of egg masses before transplanting.
- Set pheromone traps for YSB @ 20-25/ha.
- Release *Trichogramma japonicum* (Tricho-cards) against YSB @ 1 lakh/ha starting from 15 days after planting (DAP) at 7-10 days intervals 5-6 times.
- Spraying *Beauveria bassiana* product (Boverin, Biopower, Biorin) @ 1kg/ha against YSB.
- Spray *Bacillus thuringiensis* (*Bt*) based biopesticides available in the market (Dipel, Delfin, Thuricide, Bioasp, Biolep, HIL Btk) @1kg or 1lit/ha.
- Treat the seeds before sowing with fipronil (Regent) @ 25g/kg seed
- Spray need based insecticides if the pest crosses economic threshold level (ETL) i.e.5% dead heart (DH) or 1 egg mass per m^2 then apply cartap hydrochloride 4G @ 20kg/ha or fipronil 0.3G @25kg/ha or phorate 10G @ 10kg/ha or carbofuran @ 33 kg/ha in the main field. Alternately, spray cartap hydrochloride 50SP @500g/ha or fipronil(Regent) 5 SP @1 lit/ha or profenophos @ 1 lit/ha/ thiacloprid 240SP (Calypso) @ 500ml/ha or flubendiamide (39.35 SC) @175ml/ha or indoxacarb14.5 SC @ 200ml/ha (Prasad and Prasad, 2010).

Hoppers: *Nilaparvata lugens, Sogatella furcifera, Nephotettix virescens*

There are Brown and White backed plant hoppers along with Green leaf hoppers which causes damage to paddy crop. High temperature and high humidity and luxuriant growth of the plants are congenial for hoppers attack (Preap *et al.*, 2001). Plant hoppers are found in clusters at the base of the plants while nymphs and adults of leaf hoppers suck the sap from the leaves. Excess draining of the plant sap results in wilting and drying of the plants in clusters known as "hopper burn". Apart from direct feeding leaf hoppers also act as vectors of rice tungro virus disease.Management strategies for hopper control are :

- Early and synchronous transplanting would maintain population below ETL.
- Always use balance dose of fertilizers and avoid using excessive 'N' fertilizer or manure.
- Alternate wetting and drying of the paddy field also minimize the pest incidence.
- Setting up of yellow sticky traps to attract and kill hoppers.
- Foliar spraying of NSKE @5% or neem oil 0.5% or neem based commercial pesticides 300ppm @ 2.5lit/ha (Margocide OK 80%EC, Achook, Nimbicidine, Bioneem etc.).
- Keep alley place in paddy field in BPH endemic areas.
- Do not apply resurgence causing insecticides like quinalphos, chlorpyriphos, methyl parathion, deltamethrin and cypermethrin especially in endemic areas.
- When hoppers population crosses ETL (5-10 insects/hill) apply any of the following insecticides as foliar spray (thiamethoxam @100g/ha or imidacloprid@125ml/ha or clothianidin 50WDG@ 60g/ha or bifenthrin (Talstar) 10EC@250ml/ha or buprofezin 25%WP (Applaud, Buprolord) @ 300 to 500ml/ha) or Flonicamid 50WG @80gm/ha in the main field.

Gall midges: *Orseolia oryzae*

The adult female of this pest is bright red with a reddish brown abdomen. The maggots are headless, legless and remain inside the basal part of the stem. It is a serious pest of paddy in only at tillering stage. Attack of pest in growing primordia results in to hollow tubular galls known as "silver shoot" which looks like onion leaf. Affected tillers do not bear panicles. Early infestation of pest encourages profuse tillering and stunted growth of the plants. Management of gall midges includes the following:

- Avoid the mono-cropping of paddy.
- Sprouted seed can be soaked in imidacloprid 200SL @ 0.05% (2.5ml/lit) for 3 hours and then shade dried and broadcasted on raised nursery bed which will give protection against gall midge
- When the pest crosses ETL i.e. 1 silver shoot (SS)/ m^2imidacloprid spray @ 125ml/ha effective. Sprays for the management of stem borer will also give protection against gall midges.

Gundhi Bug: *Leptocorisa acuta*

It is also serious pest of rainfed paddy ecosystem during milky stage. The adult bug is slender, brownish green in colour measuring about 15mm in length with long slender legs and four segmented antennae. The nymphs are green to brown in colour. Attack at dough stage causes shrivelled grains of rice. Following practices are recommended for management of Gundi bug.

- The extract of 2.5kg garlic + 500g tobacco leaves + 500g washing powder sprayed over one hectare area during milky stage of paddy can be controls up to 80% gundhi bug incidence.
- Other indigenous technique where pest also keep in control by putting fermented snails or crabs bait @ 20-25/ha to attract and divert pests from sucking milk of rice grain found effective.
- When the pest crosses ETL i.e. 1 bug/ hill then give spray with abamectin @ 500ml/ha or phosphamidon 40SP @1 lit/ha or dusting with malathion 5%D @ 25kg/ha.

Pests of Rainfed Pulses Crop and their Management

Pulses, the food legumes, have been grown by farmers since millennia providing nutritionally balanced food to the people of India (Nene, 2006) and many other countries in the world. The major pulse crops grown under rainfed conditions are black gram, chickpea, cowpea, faba bean, green gram, horse gram, lablab bean, lentil, moth bean, pea and pigeon pea. There is nearly 250 insects have been recorded feeding on pulse crops (Singh *et al.*, 2009). Of these, about one dozen insects including Pod borers, Stem borers, Leaf miners, Foliage caterpillars, Cutworms, Jassids, Aphids and Whiteflies are most important. Some polyphagous insects also feed on these crops and cause considerable damage.

Aphids: *Aphis craccivora*

It is a greenish black coloured aphid. Sap of plant is suck by both nymphs and adults which causes pods to become deshaped, withered and malformed. Severe infestation may result in complete drying of affected pods.Management strategies for aphids:

- Bio-intensive Integrated Pest Management (IPM) measures, such as the release of commercially-reared beneficials (applied biological control) and the application of soft pesticides (soaps, oils, botanicals) can be used which will also increase native natural enemies.
- Using of entomopathogenic fungus *Beauveria bassiana* product @ 1kg/ha to cause epizootics in aphids in the pulses field.
- Spraying of infested crop with Acetamiprid 20SP @ 20gm or 125 ml imidaclorpid or Flonicamid @ 60gm in 500 L water per ha effectively control aphids. As the strong point of this pest lies in its very quick multiplication, the insecticidal treatment has to be repeated as soon as aphid population is found to have built up again.

Thrips: *Megalurothrips distalis, Caliothrips indicus*

Very small yellow fringe winged adults can be seen on flowers, leaves and or growing buds. The leaves are mottled with characteristic silvering due to the attack of insect especially under dry spell on lab lab, black gram, green gram, cow pea. Later leaves dry and shed. Damaged plants do not develop pods.Management strategies includes;

- Install yellow sticky trap to monitor and attract the thrips.
- Spray Fipronil 5% SC @ 400 ml or Spinosad 45%SC @ 75 ml or Thiomethoxame 25% WG @ 50 gm per hectare will give best management against thrips.

Whitefly: *Bemisia tabaci*

Adult is a minute insect with yellow coloured body with white waxy bloom. The damage is caused by both nymphs and adults, which are found in large numbers. They suck plant sap and lower its vitality. Severe infestation results in premature defoliation, development of sooty mould or honey dew and shedding of flowers and pods. Vector of many virus diseases in pulse crops.Management strategies for whitefly are listed below.

- Field sanitation through destroying and removing the crop residues, and weeds is an effective practice against whiteflies.
- Combination of cultural practices and need based insecticidal applications keep a check of whitefly populations.
- Repeated applications of insecticides during early and mid seasons lead to resurgence of whiteflies, and hence a highly judicious chemical application is a must. Neem oil (1%), fish oil resin soap (2.5%) and neem seed kernel extract (NSKE) 5% give effective control of whiteflies. Triazophos 40 EC @ 600 g a.i./ha or Ethion 50 EC @ 1000 g a.i./ha or Acetamiprid 20 SP @ 30-40gm or Spiromesifen @200ml or Difenthiuron 50%WP @250g/ ha are effective against whiteflies. Proper coverage of underside of leaves during the insecticidal sprays effectively reduces the whitefly population.

Pod bug: *Riptortus pedestris, Clavigralla gibbosa*

The nymphs and adults of pod bug suck the juice of unripe seeds from the green pods. In case of severe infestation the tender parts get shrivelled and later dry up. The bugs are seen clustered around on the pods.Spraying the infested crop with Fipronil 5%SC @ 400ml or Fipronil 80%WG @ 20gm or Profenophos 50%EC @ 500ml in 700-1000 L water per hectare will give effective control of pod bugs.

Spotted pod borer: *Maruca testulalis*

Female lays eggs singly on flowers, buds or pods. After hatching larva bores buds or pods and feed on seeds. The larva bores the buds, flowers or pods; infested pods and flowers are webbed together. The larva feeds on seeds. Management strategies are;

- Summer deep ploughing is effective in destroying the eggs, larvae and pupae of pests.
- Population of pest can also be suppressed to some extent by resorting clean cultivation.
- Remove the weeds in and around the field.
- Set up light trap to monitor, attract and kill the pod borer moths.
- Conserve natural enemies like *Microbracon* sp., *Bracon hebetor* and *Apanteles* sp. parasitizes the caterpillar pests.
- Spray with profenophos @ 1 lit/ha/ thiacloprid 240SP (Calypso) @ 500 ml/ha or flubendiamide (39.35 SC) @ 175 ml/ha or indoxacarb 14.5 SC @ 200 ml/ha or Chlorantraniliprole (Coragen®) @ 125 ml/ha will give effective control.

Other than these many time mites also cause damage to pulses in dry season those can be control using Fenaxaquin @ 1000 ml or Propergite @ 750 ml per hectare on need basis.

Management of Soil Borne Diseases

Several fungal and bacterial bioagents viz., *Bacillus* spp., *Pseudomonas* spp. and *Trichoderma* spp. and *Verticillium* spp., are extensively used in agriculture by the farmers for the management of soilborne diseases. Among the fungal bioagents, *Trichoderma* is a very effective biological means for plant disease management especially the soil borne. It is a free-living fungus which is common in soil and root ecosystems. It is highly interactive in root, soil and foliar environments. It reduces growth, survival or infections caused by pathogens by different mechanisms like competition, antibiosis, mycoparasitism, hyphal interactions, and enzyme secretion.

General features of *Trichoderma* on PDA

Colonies, at first transparent on media such as cornmeal dextrose agar (CMD) or white on richer media such as potato dextrose agar (PDA). Conidia typically form within 4-6 days in compact or loose tufts in shades of green or yellow or less frequently white. Yellow pigment may be secreted into the agar, especially on PDA. A characteristic sweet or 'coconut' odor is produced by some species. Conidiophores are highly branched and thus difficult to define or measure, loosely or compactly tufted, often formed in distinct concentric rings or borne along the scant aerial hyphae.

Benefits of *Trichoderma*

Disease Control: *Trichoderma* is a potent biocontrol agent and used extensively for control of soilborne diseases. It has been used successfully against several pathogenic fungi belonging to various genera, viz. *Fusarium, Rhizoctonia, Sclerotium, Scelerotinia, Pythium, Phytophthora, Macrophomina* etc.

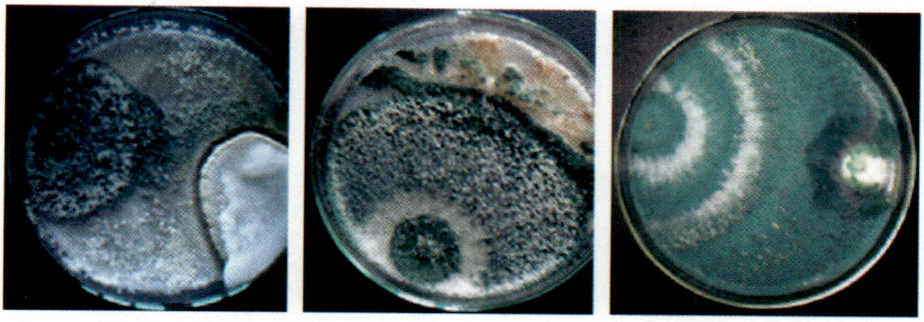

Trichoderma X S. rolfsii *Trichoderma X R. solani* *Trichoderma X Fusariumoxy sporum f.* sp. *pisi*

Plant Growth Promoter: *Trichoderma* strains have ability to solubilizing phosphates and micronutrients. The application of *Trichoderma* strains on plants increases the number of deep roots, thereby increasing the plant's ability to resist drought.

Biochemical Elicitors of Disease:*Trichoderma* strains are known to induce resistance in plants. Three classes of compounds that are produced by *Trichoderma* and induce resistance in plants are now known. These compounds induce ethylene production, hypersensitive responses and other defense related reactions in plant cultivars.

Bioremediation:*Trichoderma* strains play an important role in the bioremediation of soil that are contaminated with pesticides and herbicides. They have the ability to degrade a wide range of insecticides: organochlorines, organophosphates and carbonates as well as it has potential to scavenge heavy metals from the soil.

Biocontrol mechanisms of *Trichoderma*

The *Trichoderma* may suppress the growth of the pathogen population in the rhizosphere through competition and thus reduce disease development. It produces antibiotics and toxins such as trichothecin and a sesquiterpine, Trichodermin, which have a direct effect on other organisms. The antagonist (*Trichoderma*) hyphae either grow along the host hyphae or coil around it and secrete different lytic enzymes such as chitinase, glucanase and pectinase that are involved in the process of mycoparasitism. Examples of such interactions are *T. harzianum/T. asperellum* acting against *Fusarium oxyporum, F. roseum, F. solani, Phytophthora colocaciae* and *Sclerotium rolfsii*. In addition, *Trichoderma* Enhances yield along with quality of produce. Boost germination rate. Increase in shoot & Root length Solubilizing various insoluble forms of Phosphates Augment Nitrogen fixing. Promote healthy growth in early stages of crop. Increase Dry Matter Production substantially. Provide natural long term immunity to crops and soil.

Soybean seeds priming with Trichoderma

Pea seeds priming with Trichoderma

Method of application

1. ***Seed treatment:*** Mix 10-12 g of *Trichoderma* powder per Kg of seed before sowing.
2. ***Nursery treatment:*** Apply 10 - 25 g of *Trichoderma* powder per 100 m2 of nursery bed. Application of neem cake and FYM before treatment increases the efficacy.
3. ***Cutting and seedling root dip:*** Mix 10g of *Trichoderma* powder along with 100g of well rotten FYM per liter of water and dip the cuttings and seedlings for 10 minutes before planting.
4. ***Soil treatment:*** Apply 5 Kg of *Trichoderma* powder per hector after turning of sun hemp or dhaincha into the soil for green manuring. Or Mix 1kg of *Trichoderma* formulation in 100 kg of farmyard manure and cover it for 7 days with polythene. Sprinkle

the heap with water intermittently. Turn the mixture in every 3-4 days interval and then broadcast in the field.

5. ***Plant Treatment:*** Drench the soil near stem region with 10-12g *Trichoderma* powder mixed in a liter of water.
6. *Trichoderma* formulations.
7. Important commercial formulations are available in the name of Sanjivani, Trichoguard, Niprot, Tricha and Bioderma. These formulations contain $3x10^6$cfu/g of carrier material. Talc is used as carrier for making powder formulation.

Grain based Trichoderma

Talc based Trichoderma

Applications of *Trichoderma*

Used in Damping off caused by *Pythium* sp. *Phytophthora* sp., root rot caused by *Pellicularis filamentosa,* Seedling blight caused by *Pythium,* collar rot caused by *Sclerotium rolfsii,* Dry rot caused by *Macrophomina phaseoli,* Charcoal rot caused by *Macrophomina phaseoli,* loose smut caused by *Ustilago segetum,* Karnal bunt diseases, Black scurf caused by *Rhizoctonia* solani, Foot rots of pepper and betel vine and capsule rot of several crops. Effective against silver leaf on plum, peach & nectarine, Dutch elm disease on elm's honey fungus (*Armillaria mellea*) on a range of tree species, *Botrytis* caused by *Botrytis cinerea,* Effective against rots on a wide range of crops, caused by *Fusarium, Rhizoctonia, Pythium,* and Sclerotia forming pathogens such as *Sclerotinia.*

Recommended use of *Trichoderma*

Trichoderma is most useful for all types of plants and vegetables such as cauliflower, cotton, tobacco, soybean, sugarcane, sugarbeet, eggplant, red gram, Bengal gram, banana, tomato, chillies, potato, citrus, onion, groundnut, peas, sunflower, brinjal, coffee, tea, ginger, turmeric, pepper, betel vine, cardamom etc.

Precautions

- Don't use chemical fungicide after application of *Trichoderma* for 4-5 days.
- Don't use *Trichoderma* in dry soil. Moisture is essential for its growth and survival.
- Don't put the treated seeds in direct sun rays.
- Don't keep the treated FYM for longer duration.

Compatibility

Trichoderma is compatible with Organic manure *Trichoderma* is compatible with biofertilizers like *Rhizobium, Azospirillum, Bacillus subtilis* and *Phosphobacteria*. *Trichoderma* can be applied to seeds treated with metalaxyl or thiram but not mercurials. It can be mixed with chemical fungicides as tank mix.

Integrated Weed Management

Weeds are perceived as unwanted intruders into agro-ecosystems that compete for limited resources, reduce crop yields and quality and require large number oflabour and technology to prevent even greater crop losses. A particular plant is a weed only in terms of a human attitude. Plants considered weeds when they interfere with the utilization of land and water resources or otherwise adversely intrude upon human welfare. The most common and acceptable definition of weed is "A plant growing where it is not desired" (Terminology Committee of Weed Science Society of America, 1956). Weed Science Society of America in 1994 again defined weed as "any plant that is objectionable or interferes with the activities or welfare of man". Weeds are the major deterrent to the development of more sustainable agriculture systems. They cause enormous losses and sufferings to human beings by way of reduction in crop yield and quality, wastage of human energy and resources and increased expenditure to alleviate the problems caused by them. Despite the progress made in weed science over the last 40 years, weeds remain one of the major constraints to agricultural production in developed and developing countries. When improved agricultural technologies are adopted, efficient weed management becomes even more important, otherwise the weeds rather than the crops benefit from the costly inputs.

Seed Production Capacity and Dormancy of Weed seeds

Weed seeds remain viable for longer period without losing their viability. The seed production capacity and length of seed survival in soil of few important weeds have been given in Table.

Table 14.1 Seed production and length of seed survival of some weeds in soil (Wilson, 1988)

Weed species	No. of seed produced/ plant	Length of seed survival in undisturbed soil (years)
Ball mustard (*Neslia paniculata*)	490	10
Canada thistle (*Cirsium arvense*)	680/stem	21
Common lambsquarters (*Chenopodium album*)	72450	39
Common mullein (*Verbascum thapsus*)	223200	39
Common purslane (*Portulaca oleracea*)	52300	30
Common ragweed (*Ambrosia artemisiifolia*)	3380	39
Common sunflower (*Helianthus annuus*)	7200	1
Curly dock (*Rumex crispus*)	29500	39
Curlycupgumweed (*Grindelia squarrosa*)	29700	10
Dandelion (*Taraxacum officinale*)	15000	6
Devils beggarticks (*Bidens frondosa*)	7000	10
Green foxtail (*Setaria viridis*)	34000	39
Redrot pigweed (*Amaranthus retroflexus*)	117400	10
Wild oat (*Avena fatua*)	250	1
Yellow foxtail (*Setaria glauca*)	6420	30

Crop-weed Competition

Competition is the rivalry between crop and weed for a common demand. Competition does not occur until the immediate supply falls below the combined demand of both crop and weeds. *Competition is what weed control is all about*. Weeds compete with the crop for nutrients, moisture, solar radiation and space when they are limited and reduce the crop yield and quality. Crop-weed competition is largely affected by nature and intensity of weeds, crop species and cultivar, light, moisture, nutrients and management practices.

*Competition for light:*Weeds compete with the crops for light by faster and higher growth, large leaves and climbing devices. The competition for light begins when plants begin to shade each other. Cudney*et al.*, (1991) showed that wild oat (*Avena fatua*) reduced light penetration and growth in mixture with wheat by producing greater height growth than wheat. Similarly in soybean, velvetleaf (*Abutilon theophrasti*) intercepted more light than soybean due to its greater height growth and dry matter allocation to more branches in the upper layers of the canopy (Akey *et al.*, 1990).

Competition for nutrients: Nitrogen, phosphorus and potassium are the primary plant nutrients. In general, weeds remove more nutrients than crops. Uncontrolled weeds in any crop remove more quantity of nitrogen and potassium than the phosphorus. Fertilization usually stimulates weed growth to the crop's detriment. With low fertility, competition is usually for nutrients, however with high fertility, competition is just as vigorous, and primarily for light.

Competition for water: Competition for water between weeds and crops reduces soil moisture availability and causes water stress. Weeds are comparatively more tolerant than the crop plants to moisture stress. In general, for producing equal dry matter, weeds transpire more water than do most of our crop plants. Shahi (1978) determined the consumptive use of water of *Chenopodium album* as 550 mm against 479 mm for wheat. It was further noted that the weeds removed moisture evenly from up to 90 cm soil depth, while the major uptake of moisture by wheat was limited to the top 15 cm of soil.

Weeds and Water-use efficiency (WUE)

Water is the greatest limiting resource for crops and its management is one of the biggest challenges in agriculture. In the field, the cultivated plants and weeds take share in influencing the water balance.Agriculture is responsible for 70% of all water use globally and water use efficiency (WUE) in this sector is low, not exceeding 45% (Zhang *et al.* 2004).The annual freshwater withdrawals for agriculture in 2001 amounted to 83 percent.Weeds compete for water, cut water availability, and contribute to crop water stress. Knowledge of weed transpiration (T) is important in assessing the competition of weeds against cultivated plants.

Aquatic weeds have the potential to use excessive quantities of water through extensive root systems and high transpiration rates. Weeds present in the canals and ditches also can obstruct water flow. Reducing flow rate caused by excessive growth of submerged weeds was determined by 80% in some small canals (El-Shinnawy*et al.* 2000).

Table 14.2 Transpiration ratio (T: R1) of various crops and weed species (Croissant *et al.*, 2014)

Crops	T:R1	Weeds	T:R1
Sorghum	304	Pigweed	287
Corn	349	Lambs quarters	801
Cotton	568	Gum weed	606
Sunflower	630	Ragweed	948
Sugar beet	377		
Soybean	646		
Wheat	528		
Dry beans	700		

T:R1: Pounds of water transpired per pound of above-ground dry matter produced. Water weighs 8.34 pounds/gallon.

In Egypt the total water loss by ET from water hyacinth infested areas was estimated to be 3.5 billion m^3 per year. This amount is enough to irrigate about a further 432 km^2 (43200 ha) every year (Fayad *et al.*, 2000). Water hyacinth causes 4 billion m^3 losses of water every year in Egypt, enough to sustain Cairo with water (El-Shahawy, 2013).It was estimated that a pond infested with one hectare of water hyacinth will produce up to 1.8 tons of dry mass a day. The aquatic weeds pose a big problem in water loss because it have higher transpiration rate. Several recent studies have shown that such water losses are 2, 3 or even 6 times higher in reservoirs covered in weeds than they are in open waters (Penfoundand Earle, 1948). The water loss (evapotranspiration) caused by water hyacinth weed was estimated by about to be 2.5 and 13 times evaporation from that of a free water surface and the flow of water in canals is reduced drastically was 40 to 90%.

Water Loss by Weeds

Over the coming decades, global climate change will affect weeds. Fourteen of the world's worst weeds are C_4 plants, while 76% of the harvested crop area is C_3 plants. In drought situations C_4 weeds might also have advantages over C_3 crops under elevated CO_2. Elevated CO_2 increase plant growth (above-and belowground) and improve plant water relations (reduces transpiration and increases WUE). Weeds often show greater growth responses to elevated CO_2 than do crop plants, which may be the result of weeds having greater genetic diversity and physiological plasticity than managed plants.

Critical period of crop-weed competition

Weeds are capable of accumulating dry matter faster than the crop plants. Thus, the duration of weed infestation and time of weed removal has a significant influence on crop growth and economic yield. A fundamental principle of plant competition is that early occupants on a soil tend to exclude the later ones. Removing weeds at any time during growing season may not be beneficial. Stage of weed removal is as important as removal *per se*. Therefore, it is necessary to identify the critical period of crop-weed competition to render the weed control practices more effectively. It can be defined, as "the shortest span of time in the ontogeny of crop growth when weeding will result in higher economic returns". Mishra (1997) reviewed the research work done under AICRP-Weed Control and elsewhere since last 20 years and reported that the critical period varies from 15- 30 days after sowing (DAS) in greengram, blackgram, cowpea and okra; 15-45 DAS in direct seeded rice, maize, sorghum, sunflower, sesame and rapeseed-mustard; 30-45 DAS in transplanted rice, pearl millet, wheat, peas, jute, cauliflower, cabbage and tomato; 20-45 DAS in soybean and potato; 30-60 DAS in castor, lentil, chickpea, frenchbean, and groundnut; 15-60 DAS in pigeonpea and cotton; 30-75 DAS in onion and 30-120 DAS in sugarcane depending upon nature and intensity of weed flora, agro-climatic situations and management practices.

Losses due to weeds

Weeds are major agricultural pests that can devastate a crop if not properly managed. Unlike insect pests and pathogens, where the effect on crop is visible, the damage done by weeds is often noticed. Losses caused by weeds in the developing world are still at least 37% annually and in certain regions, losses can be up to 90-95% including total crop failure. In addition to direct effect on crop yield, weeds result in considerable reduction in the efficiency of inputs used. The precious and costly inputs such as fertilizers and irrigation water which are otherwise meant for realizing the potential yield will be usurped by the weeds.

In addition to their direct and indirect effects upon India's economy, invasive weeds are recognised as the biggest threat to India's biodiversity next only to habitat destruction. Estimates on potential and actual losses despite the current crop protection practices are given in Table (Oerke, 2006). The total global potential loss due to weeds varied from about 23% in wheat to 40% in maize. In India, weeds produce the highest loss (33%) followed by pathogens (26%), insects (20%), storage pests (7%), rodents (6%) and others (8%). Based on the average loss (20%) due to pests as estimated by Ministry of Agriculture, and present value of our agriculture produce as Rs. 7 lakh crores, it approximates to around Rs. 1, 40, 000/- crores, of which weeds are responsible for more than 46,000 crores (Agarwal, 2007). The precious and costly inputs such as fertilizers and irrigation water which are otherwise meant for realizing the potential yield will be usurped by the weeds. Indirectly, weeds also act as an alternate host for many crop pests and also some of them like *Parthenium* cause serious health hazards to human beings.

Table 14.3 Critical period of crop-weed competition and losses due to weeds in different field crops (Mishra, 1997).

Crops	Critical period	Average yield reduction (%)
Cereals		
Rice (Direct seeded)	15-45	50-90
Rice (Transplanted)	30-45	15-40
Maize	15-45	40-60
Sorghum	15-45	15-40
Pearlmillet	30-45	15-60
Wheat	30-45	20-40
Pulses		
Pigeonpea	15-60	20-40
Greengram	15-30	30-60
Blackgram	15-30	30-50
Cowpea	15-30	30-50
Chickpea	30-60	15-25
Peas	30-45	20-30
Lentil	30-60	20-30

Oilseeds		
Soybean	20-45	40-60
Groundnut	40-60	40-50
Sunflower	30-45	30-50
Castor	30-60	30-35
Safflower	15-45	35-60
Asesamum	15-45	15-40
Rapeseed-mustard	15-40	15-30
Linseed	20-45	30-40
Commercial Crops		
Sugargane	30-120	20-30
Cotton	15-60	40-50
Jute	30-45	50-80
Vegetables		
Potato	20-40	30-60
Cauliflower	30-45	50-60
Cabbage	30-45	50-60
Okra	15-30	40-50
Tomato	30-45	40-70
Onion	30-75	60-70

Integrated Weed Management System (IWMS): Concepts and Components

Integrated weed management (IWM) is a component of integrated crop management (ICM) which a transient all disciplines. IWM involves three elements;

a) Multiple tactics (e.g., competitive varieties, cultural practices and herbicide usage).

b) Weed population maintained at levels below that cause economic damage, and

c) Conservation of environmental quality.

Buchanan (1976) defined IWMS as "the application of many kinds of technologies in a mutually supportive manner. It involves the deliberate selection, integration and implementation of effective weed control measures with due consideration of economic, ecological and sociological consequences". The research approach to the development of an IWMS must take all aspects of the cropping system into consideration. Invariably, each cultural practice influences the competitive ability of both the crop and the weed community leading to a multitude of complex interactions. However, efforts must be made to work within the existing production practice to ensure a greater likelihood of acceptance by the farming community. Although every agronomic practice starting from planting to harvesting of crop has implications on weed competition, farmers are neither aware of it nor appreciate it. It is important to educate farmers about this. The effect may be small for individual practice but effect would be substantial if all such practices are combined effectively. IWMS is a knowledge intensive exercise where scientists may also be discouraged to get involved. In India, at least we are yet to generate information on economic threshold limits (ETL) for different weeds, weed seed dynamics in soil, understanding of eco-physiology of crop-weed competition, weed flora shifts in crops and cropping systems etc. It is time we started working on these. Besides developing effective IWM packages for different crops and cropping systems, it is very important to demonstrate them to farmers on large scale.

Components of IWMS

The weed management in field crops requires an integrated approach that utilizes effective cultural, mechanical, biological, ecological and chemical methods in a mutually supported manner into the crop production system with due consideration of economic, environmental and sociological consequences. Various methods of weed management such as prevention, cultural, mechanical, chemical etc. have been used with different degree of success in different agro-ecological zones and productions system, which are as follows.

Weed Prevention: Prevention involves procedures that inhibit or delay weed establishment in areas that are not already inhabited by them. It encompasses all measures taken to forestall the introduction and spread of weeds. Success of a preventive programme varies with species and the amount of effort devoted to control. The prevention of a weed problem is usually easier and less costly than control or eradication. The following measures can be suggested to prevent the introduction of weeds into non-inhabited field.

- Use 'clean' (weed seed-free) crop seed for planting.

- Use organic manures only after thorough decomposition to kill weed seeds.
- Clean harvesters and tillage implements before moving to non-weed infested area.
- Avoid transportation or use of soil from weed infested area.
- Inspect nursery stock or transplant for seed and vegetative propagules of weeds.
- Remove weeds that are near irrigation ditches, fence rows, right-of-way and other non-crop land.
- Prevent reproduction of weeds.
- Use weed seed screens to filter irrigation water.
- Restrict live stock movement in to non-weed infested area.

Other practices used to prevent and avoid potential weed problem at the state, regional or national level are weed laws, seed laws and quarantines.

Cultural Methods

Cultural weed management is referred to as the use of any methods that directly enhance crop competitive ability against weeds. Despite the great progress made in agriculture, manual and mechanical methods continue to be important weed management practices in many regions of the world. Cultural methods are used to complement manual and mechanical methods. Cultural practices are manipulated in such a way that they become more favourable for crop growth and less to weeds. They are not only eco-friendly but also reduce the use of costly herbicides.

In *organic agriculture*, the approach to weed management involves the whole cropping system. The aim is not the eradication of weeds but to maintain a balance between crop plants and the weeds, with the grower adjusting the balance in favor of the crops whenever possible. Cultural practices that shift the balance of competition towards the crop usually will disfavor weed occurrence and improve crop yields. Crop rotation is at the heart of the organic system and although the method and timing of soil cultivations and choice of crop are usually linked, both contribute in different ways to manipulation of weed population. Any practices that provide vigorous uniform crop establishment usually will assist in reducing weed prevalence. Mechanical Method.

Weed removal in India is largely restricted to mechanical and cultural methods. Mechanical weed control involves removal of weeds with various tools and implements including manual removal by hand weeding. Inter-cultural operations are performed primarily to destroy the weeds present in the field and create favourable soil conditions for crop growth. Hand hoeing and manual weedings are most common practices performed for weed control in almost all pulse crops. One or two hand weedings at critical crop-weed stage provide satisfactory control of weeds in most of the pulse. Extra weeding may be required if crops are infested with perennial weeds. The use of mechanical weed control demands experience and knowledge of how to use the soil as a weed control agent. Climate and soil type play an important role in the possibilities for mechanical weed control. Monitoring the early development of weeds is necessary for timing of weed harrowing at the optimum stage. Mechanical weeders include cultivating tools such as hoes, harrows, times brush weeders, and cutting tools like mowers. The success of mechanical weeding depends upon the stage of weeds, crop geometry and climatic conditions. Hand weeding may also be used after mechanical inter row weeding to deal with weeds left in crop rows.

Biological Control

Biological control of weeds is the deliberate use of natural enemies, primarily insects or fungi, to suppress growth or reduce the population of weed species. The utilization of biological weed control agents in weed management offers many advantages including,

a) A high degree of specificity for the target weed.
b) No effects on non-target and beneficial plants or man.
c) Absence of weed resistance development.
d) Absence of residue builds up in the environment.

The two most common approaches are; the classical or inoculative method and the bioherbicide or inundative strategy. The classical strategy is an ecological approach, which relies on the ability of an organism to multiply and spread following small-scale release. The organism then remains in balance with the target weed, keeping the latter at an acceptably low level.

The bioherbicide strategy, the second major strategy of biological control of weeds, is to attack a whole weed population with a single, non-persistent application of inoculum. The inoculum inundates the whole of the area infested by a weed to increase the effectiveness of the agent. Generally, the pathogens considered for this strategy are fungi, hence the commonly used name-mycoherbicides. *Colletotrichum gloeosporioides* f.sp. *aeschynomene* (Collego) and *Phytophthora palmivora* (DeVine) are currently being treated as mycoherbicide for the control of northern jointvetch (*Aeschynomene virginica*) and Strangler vine or milkweed (*Morrenia odorata*).

The use of insect pests as bio-control agents has application in situations where only one weed is a problem such as in aquatic ecosystem, range lands, forestry or waste lands. The control of *Parthenium hysterophorous* with Mexican beetle *Zygogramma bicolorata* and the control of water hyacinth with weevil *Neochetina spp.* are a couple of successful examples of use insect pests for weed control. The Mexican beetle is very effective on *Parthenium* and excellent control of the weed has been reported from several places in India. However, the insect is active only during the rainy season and it hibernates in soil during the dry and winter months.

Chemical Method

Weed control through manual/mechanical though very effective, has certain limitations such as unavailability of labour during peak period, high labour cost, unfavourable environment particularly in rainy season etc. Under such conditions, use of herbicides is advantageous and economical.

In general, herbicides account for the largest proportion of crop protection chemicals sold on a world-wide scale. Globally, the herbicides constitute 45% of the total pesticides sale and in some countries like the USA, Germany and Australia; the figure is as high as 60-70 %. In India, however, the herbicides form a meager 16 % of total pesticide consumption. Sharp increase in wages and unavailability of labour due to industrialization, urbanization and Government schemes like MNREGA are bound to make herbicides more acceptable to farmers. Currently about 95 % of the herbicides is consumed in wheat (42%), rice (30%) and tea plantation (23%). The use pattern might change soon with the introduction and indigenization of more chemicals.

In India, the chemical weed control is mostly prevalent in states of Punjab, Haryana, Western UP, Tamil Nadu and Andhra Pradesh. In Punjab and Haryana, the major use of herbicides is in rice and wheat crop, Tamil Nadu and Andhra Pradesh (in rice), Madhya Pradesh (in soybean and wheat) and Gujarat (in groundnut and spices). Nearly 95% of rice and 85% of wheat in Punjab and 60% of rice and 40% of wheat in Haryana is using new technology based on herbicide use for weed management.

Advantages of herbicides

- Herbicides can kill many weeds that survive by mimicry, e.g. *Phalaris minor* and *Avena* spp. in wheat and *Echinochloa* spp. in rice.
- Herbicides can be safely used where manual weeding damages the crop, e.g. onion, garlic, groundnut, etc.
- Herbicides are more effective on perennial and parasitic weeds.
- Herbicides can effectively be used in closely planted crops where manual and mechanical weeding is not possible.
- Herbicides are safe on erodable and sloppy lands where tillage may accelerate soil and water erosion.
- Herbicides enable early and timely planting of crops.
- Herbicides can be used where physical condition of soil is not conducive to manual weeding, e.g., during rainy spells.
- While one time application of a herbicide is often enough, the manual/mechanical method is repetitive.
- Herbicides kill weeds *in situ* without permitting their dissemination.
- Herbicides can safely be used to control weeds growing in obstructed situations such as right-of-way, under fruit trees and on undulating lands.
- Herbicides provide timely weed control by controlling weeds at the critical period of their competition.
- In dry land agriculture, effective herbicidal control ensures higher water use by the crops and less crop failure due to drought.

- Planning and management of labor for various agricultural operations in big farms are facilitated by use of herbicides.
- Use of herbicides offers gender equality as manual weeding is largely done by female laborers.
- In many parts of the country the cost of weed control through herbicides is lower than that of manual weeding.

Concerns

Despite the above advantages, several concerns such as weed shift, toxicity and residue problems, herbicide leaching, and adverse effect on non-target organisms like fishes, beneficial microorganisms and appearance of herbicide resistant weeds have also been made with the indiscriminate use of herbicides.

Precautions in herbicides use

- Hand weeding may be given wherever necessary in addition to herbicide treatment.
- Use lower rates in light soils and higher rates in higher soils.
- These recommendations are based on the sole cropping.For herbicide use in inter-cropping system, consult experts.
- Apply herbicide using a sprayer with 500-600 lit/ha water.
- Also spray solutions immediately after preparation. If delay is inevitable, stir the solution or shake the sprayer thoroughly before application to prevent settling down of herbicide in the bottom.
- Flood jet or flat pan nozzle should be used for spraying the herbicide.
- Match the swath correctly. Overlapping of spray may lead to crop phytotoxicity and gaps may lead to poor control of weeds.
- Separate sprayer need to be used for herbicide and for insecticide spray. If not, then sprayer has to be thoroughly washed with soap water after the herbicide spray before it is used for insecticide spray.
- Avoid spraying on windy days as drifting may damage the neighbouring susceptible crops.

- Do not spray against the wind.
- Avoid smoking or eating during spray application.
- Wash sprayer thoroughly after each spray.
- Handle chemicals carefully. Always keep them away from children and food articles.
- Dispose the empty container by burying in the soil or by burning.
- Use protective clothing (gum boots, gloves, etc.) goggles and mask wherever possible.

❑❑❑

15

Soil Health Management

Acid Soil Management

Acid soils occupy 90 Mha of land constituting over 1/4th of the total geographical area (TGA) of India (Table 13.1). About half of such area is under cultivation and rest is under forestry and other uses. About 25 mha of cultivated lands with pH below 5.5 are critically degraded with very poor physical, chemical and biological characteristics and thus are of poor soil health. Highly acidic soils in India are restricted to the Himalayan region, red and lateritic region (comprising both the southern and eastern plateau) and some pockets in the greater plains of the country. The soils with pH < 4.5 covers 1.88 % of the TGA measuring about 6.2 Mha area. Moderately acidic soils (pH 4.5 to 5.5) cover an area of 24.7 Mha or 7.56% of TGA.Soil acidity is a result of acid build-up in the soil. The production of acid in the soils is a natural process and many soils in the high rainfall areas are inherently acidic.

Major reasons for Soil Acidification

- Rainfall and leaching (basic ions are removed through leaching).
- Acidic parent material such as Granite.
- Organic matter decay (organic acids are produced).
- Harvest of high yielding crops and removal of bases.
- Hilly terrain and loss of alkali and alkali earth metals.
- Persistent use of acid forming nitrogeneous fertilizers.

- The normal agricultural practices of fertilising and harvesting, along with increases in leaching of nitrate nitrogen and occasional soil erosion, accelerate the acidification of agricultural lands.

Acid soils of Jharkhand

The soils of Jharkhand state belongs to the order Alfisols and 80% of the soils are acidic in reaction. The soils are sandy loam to loamy texture in nature. The major constraints in the area are steep slopes with undulating topography, severe water erosion and low available water capacity. Erratic rainfall, lack of irrigation facilities, poor water retentive capacity and permeability of the soils are the major problems limiting successful double-cropping. The pH of the soil varied from 4.5 to 5.5 in the upland to low land situations. The organic carbon content varied from 0.35-0.45 % in the upland to low land situations.

Table 15.1 Extent and distribution of acid soils in different states of India (Mha)

States	Strongly acidic (pH < 4.5)	Moderately acidic (pH 4.5-5.5)	Slightly acidic (pH 5.5-6.5)	Total	% TGA
Assam	0.02	2.31	2.33	4.66	59.41
Bihar	-	0.04	2.32	2.36	25.06
Jharkhand	-	1.00	5.77	6.77	84.93
Odisha	-	0.26	8.41	8.67	55.68
West Bengal	-	0.56	4.20	4.76	53.63
Chhattisgarh	0.15	6.30	4.49	10.84	80.18
Goa	-	0.11	0.19	0.30	81.04
Himachal Pradesh	-	0.16	1.62	1.78	31.97
Jammu & Kashmir	-	0.09	1.48	1.57	7.06
Karnataka	-	0.06	3.25	3.31	17.26
Kerala	0.14	2.87	0.75	3.76	96.75
Madhya Pradesh	-	1.12	10.60	11.72	38.02
Maharashtra	-	0.21	4.33	0.54	1.75
Arunachal Pradesh	4.78	1.74	0.27	6.79	81.08
Manipur	-	1.44	0.32	2.19	98.09
Meghalaya	-	1.19	1.05	2.24	99.87
Mizoram	-	1.27	0.78	2.05	97.24
Nagaland	0.12	1.48	0.05	1.65	99.52
Sikkim	0.28	0.32	-	0.60	84.55
Tripura	0.06	0.75	0.24	1.05	100
Tamil Nadu	0.21	0.35	4.29	4.85	37.29
Uttarakhand	-	1.18	2.30	3.48	65.07
Total	6.19	24.81	58.94	89.94	

Source: NBSSLUP (ICAR), Nagpur

The available N and K are low to medium status throughout the entire soils. The availability of P is very low due to chemical fixation by sequioxide. There is a deficiency of Ca and Mg in the surface soils and presence of Ca-concretions in the subsurface horizons due to leaching of bases during rainy season and subsequent deposition in the lower layer. Increased solubility and toxicity of Mn and Fe, reduced availability of boron and molybdenum is prevalent in the soils.

Major characteristics of acid soils of India

1. Low soil pH (<5.5) is associated with a number of soil chemical and biological characteristics that adversely affect plant growth and soil health.
2. Exchangeable aluminum Al^{3+} concentrations increases especially in soils below pH 5.0. Excess Al^{3+} is toxic to plants and severely restricts root development.
3. Molybdenum deficiency especially when pH < 5.6.
4. Survival of bacteria is less in acid soils as compared to fungi. Thus activity of useful bacteria (N-fixers etc) is hampered.
5. Legume nodulation and biological nitrogen fixation are adversely affected by Mo and P deficiency which reduce growth of host plant.
6. Reduced base saturation of especially exchangeable Ca and Mg in soil. Low levels of Ca and Mg can also cause livestock health problem such as milk fence and grass tetany.
7. Strong P adsorption in soils due to presence of kaolinitic clay minerals.
8. In some soils, manganese toxicity develops at pH 5.0. but it is not commonly encountered.

Acid soils occurring in different climatic regions are classified as follows:

1. Acid soils of temperate climate including podzol, brown podzol, grey brown podzol, brown forest soils, and grey forest soils.
2. Acid soil of tropical and subtropical climates including yellow podzolic soil, lateritic soil and latosols.

3. Acid soils of other great soil groups including wet soils (hydromorphic soils), washed peaty soils, mucky, cat-clay (acid sulphate) soils. Cat-clay or acid sulphate soils with pH 3.5 or lower are the soils that abound in organic material as well as H_2SO_4.

In modem system of soil classification, acid sulphate soils have been assigned separate position and these soils have been placed in a group called sulpha-aquepts. This group includes soils in which the top horizon contains sulphuric horizon at some level or the other in top 25 cm thick layer. This is mineral or organic sublayer with yellow colouration due to xarocite.Acid sulphate soils of tropics possibly belong to Typic Sulpha-aquepts and those of temperate regions are mainly Typic Sulpha-aquepts and Haplaaquepts. Mineral and organic soils rich in sulphur which remain regularly submerged are referred to as sulphidic soils. Since such soils develop under the influence of saline water which is rich in sulphur, they are placed in Halic subgroup.

Importance of liming acid soils

The most important benefit of liming acid soils is reduction in solubility of the potentially toxic elements hydrogen, aluminium (Al^{3+}) and manganese (Mn^{2+}). The hydrogen ion concentration increases as pH drops below 7.0, but hydrogen only becomes toxic to plants in extremely acid soils (pH<4.0) and at very low calcium levels. As pH drops below 5.5, the concentration of soluble aluminum increases and becomes toxic to plant root growth. Below pH 5.2, the concentration of manganese also increases and can become toxic.

Soil acidity and plant growth

Soil acidity is important because it determines solubility of soil minerals and it affects many microbial processes such as organic matter decomposition and nitrogen fixation. Plant nutrient availability is greatly influenced by soil pH. With exception of P, which is most available within a pH range of 6 to 7, macronutrients (N, K, Ca, Mg and S) are more available within a pH range of 6.5-8.0, while the majority of micronutrients (B, Vu, Fe, Mn, Ni and Zn) are more available within a pH range of 5 to 7. Outside these optimal pH ranges, nutrients are available to plants in lesser amounts.

Physical

The upland acid soils are coarse textured, usually compact at surface and have low organic matter, besides distinct acidity. Low water holding capacity, high infiltration rate, high permeability, higher bulk density and soil crusting are general features of these soils affecting seed germination, soil and nutrient losses.

Restoring organic matter status through organic addition, addition of fine earth fraction, and mulches, ridge and furrow tillage, conservation tillage, contour and strip cropping, bench terracing, raising of field bunds for in situ rain water harvesting, intercropping of cereals with legumes are some of the appropriate measures for raising crop productivity in uplands.

Chemical

Common problems in acid soils in respect of chemical properties are, (1) Low pH, (II) Low CEC (due to dominance of 1:1 type clay minerals), (iii) low base saturation (16-67%), (iv) high Al, Fe, Mn saturation percentage, (v) High P-fixing capacity. These problems could be managed by amelioration by liming which improves pH, base status, CEC, inactivates Al, Fe and Mn in soil solution, reduces acidity and P-fixation in soil.

Microbial

Soil microorganism activity is greatest near neutral conditions, but pH ranges vary for each type of microorganisms. Acidic soils are one of the stressed environments for microorganisms and they are normally absent at pH values below 4 or 5.

Management Strategies in Acid soils

Liming: Liming is widely accepted practice for ameliorating acid soils. For acceptability of liming practice by farmers, it should be cost effective and the recommendation should be made in small doses. The speed at which limestone reacts in a soil to neutralize acidity, supply calcium and magnesium, and affect crop growth is largely determined by particle size. Smaller particles have more specific areas to contact soil acidity, thereby producing more rapid change in pH. The amount of limestone needed to adjust soil pH to ma desired level for a particular crop depends on the initial soil pH, soil clay content and soil buffer capacity resistance to pH change. Lime placement is important in determining reactivity and movements into the subsoil. Broadcast application is almost always recommended and where possible, the lime should be incorporated in the soil as deeply as possible by plough. Crops differ in their response to lime application.

1. High response group (pigeon pea, soybean, cotton).
2. Medium response group (gram, lentil, groundnut, maize, sorghum, wheat, pea).
3. Low response group (barley, minor millet and paddy).

For a soil with sub-soil acidity, gypsum or phosphogypsum is useful. Ca from surface applied gypsum moves down the profile much faster than lime and corrects sub soil acidity. Gypsum can be a source of S for surface soil, but it cannot amend the surface acidity.

- Applying lime @ 25% of lime requirement in alternate years is as good as lime application in full dose.
- Application of lime @ one tenth of lime requirement/ha in furrows along with fertilizers and FYM improved crop yield by 14-52% in case of crops like arhar, cotton, maize, groundnut, gram, pea, wheat etc.

Organic matter: Organic matter additions as crop residues, and/ or green manure crops in rotation or as intercrops must be the key components of crop management systems for acid upland soils planted to food crops. They should be left on the soil surface as a mulch, incorporated into the soil, or used both ways. The complexation of Al in soil solution by decomposition product of organic materials, particularly low molecular weight organic acids. The mechanism that best explained the release and subsequent hydrolysis of calcium ions. The hydroxyl ions released in the hydrolytic reaction then reacts with both the exchangeable hydrogen and aluminium ions to form water and insoluble aluminium hydroxide, respectively. Organic matter incorporated into soil contains a substantial amount of Ca that leaches and moves down the soil profile and checks Al toxicity. By this Ca and pH levels are raised in both surface and sub-surface soils.

Flooding: In lowlands systems, flooding may be an effective technique in raising the pH of the soil. However, this effect is only good for the time for which the soil is flooded. Flooded or paddy mineral soils are 'self-liming'. When they are flooded and become anaerobic (lack of oxygen in the soil atmosphere) for a period of time, the pH rises toward neutrality even when the soil pH was originally acidic. If the soil is subsequently drained and becomes more aerobic (more oxygen in the soil atmosphere), the pH will return to an acidic state

- However, care must be taken if the soil contains manganese-oxide minerals, since flooding conditions may lead to manganese toxicity.
- Crop consideration is also required. Flooding conditions reduces the oxygen within the soil, which is needed for plant life. As a result, crops that do not tolerate high amounts of water and low oxygen levels are not be suited for flooded conditions. Taro and rice are examples of crops that grow well in flooded lowlands.

Nutrient management in acid soils

Nitrogen: High leaching losses of applied N greatly reduce the efficacy of urea in light textured Alfisols of hill and plateau region. To overcome this problem, split application of N, neem extract coated urea, cluster bean gum polymer coated urea, shell lac coated urea and urea super granules applied in 2 split was superior.

In tomato, delayed basal N application by 10 days after transplantation and increased number of split applications from 2 to 3 resulted in highest yield of 'ArkaVikas' by applying 75 % of recommended N dose. In hybrid, chilli, application of 3 equal splits (basal + 2 top dressings) at 100 % recommended dose was the best provided the basal dose was applied at the time of transplantation owing to its quick growth habit. The relative proportion of N in the first split application when the crop is still young is reduced to 33% from 50% which improves the use efficiency of N applied.

Phosphorus: Phosphorus is relatively immobile nutrient in soil. When it reacts with the soil it tends to become "fixed" into less soluble forms with Al^{3+} and Fe^{3+} prevalent in acid soils. To avoid this phosphatic fertilizer should be banded at an appropriate depth, in close proximity of the zone of high density of roots, to attain higher Putilization. Accordingly banding of superphosphate at 5 cm depth in French bean, brinjal, tomato, onion and chilli was found to be superior (Shallow rooted crop). A saving of 40 % in brinjal and 20 % in tomato and onion was possible without any loss of yield. In deep rooted okra crop, a deeper placement at 10 cm depth was necessary to attain high P utilization.Prevalence of Fe and Mn in acid soils pre-dispose soluble form of P present in phosphatic fertilizers to less soluble available precipitates. In recent times, DAP has practically replaced SSP due to cost of raw material. Over the last two decades, this has led to increased soil acidity, undesirable build up of P leading to an induced Zn deficiency as also the emergence of S deficiency in many crops. There is an immediate need to evaluate the effect of indiscriminate use of DAP on the soils of different areas and re-induct SSP to contain these problems.

Potassium: The predominant leaching regime in acid soils leads to deficiency of Ca and Mg in soil solution. The status of K as also the balance of Ca and Mg vis-à-vis K becomes more fragile in coarse red soils than in heavier red soils having higher clay content.

Calcium, magnesium and sulphur: Most of the acid soils are deficient in Ca, Mg and S except acid sulphate soils which contain high amount of S. soils having Ca saturation less than 25 % of the total CEC requires Ca application to most of the crops. Deficiency of Ca and Mg can be corrected by using lime or liming material @ 4-5 q/ha. Coarse textured acidic upland soils low in organic matter show S-deficiency. Al-toxicity can induce Mg deficiency, liming with dolomitic limestone may be beneficial by raising pH and reducing exchangeable Al concentrations.

*Boron and molybdenum:*Acid conditions pre-dispose soils to the deficiency of B and Mo. B deficiency is the foremost (being low in available B having hot water soluble B of 0.1 µg/g) because of crop losses due to brown rot in cole crops and low yield of most vegetable crops in acid soils of the plateau. Deficiency of B caused severe brown rot that led to heavy crop losses of 70-90 % in kharif and 30-35% in rabi season in Jharkhand plateau. Application of B @ 1.5 kg B/ha to soil or 3 foliar sprays of 0.125% boric acid in cauliflower boosted curd yield by 40-95 % and reduced crop losses to 5-20 %. The losses due to Mo deficiency were 10-16 % during rainy season and 1-2 % during winter. Availability of Mo can be improved by liming. In Jharkhand soils, soil application of Mo at 1.5 kg/ha was effective in overcoming whiptail malady and boosting curd yields for 3 successive cauliflower crops. Foliar spray of 0.0125% ammonium molybdate with 0.1% surfactant or a humectants like $CaCl_2$ (0.2%), applied thrice, similar to boric acid sprays, was effective. It was necessary to apply all the foliar sprays before curd initiation to avoid undesirable green coloration of the curds due to Mo sprays. Application of B and Mo together either to soil or to the foliage showed a synergistic effect on curd yield.

Zinc: Deficiency of Zn is universal in Indian soils. The problem has become more severe due to intensive cultivation of crops without adequate augmentation of Zn. Foliar application of Zn is easy and convenient to apply, effective and cheap. It is useful to adopt un-neutralized, low concentration foliar spray (< 0.25%) in preference to neutralized 0.5% aqueous $ZnSO_4$ as the latter is practically devoid of active Zn due to precipitation. Combined application of 0.25 % $ZnSO_4$ and 0.1% boric acid, thrice enhanced the yield and superior than multi nutrient foliar sprays.

Iron toxicity: Iron toxicity in red and lateritic soils is a problem when the dissolved Fe in the rooting medium exceeds 300-500 ppm. Iron toxicity in plants promotes sterility and the grain yield is reduced by 20-80 % depending upon the situation. Iron toxicity in low land rice can be corrected by providing intermittent drainage. Application of high dose of K helps in creating oxidizing zone around rice rhizosphere grown under Fe-toxic condition, thereby preventing reduction of Fe^{3+} to Fe^{2+}. This ultimately restricts the Fe uptake by rice and the yield is improved. Liming had beneficial effect on alleviating Fe-toxicity in rice.

Manganese toxicity: Many acid soils of India contain Mn containing minerals. When the soil pH drops below 5.5, Mn minerals become highly soluble and become toxic. When these minerals dissolve, Mn ions are released into soil solution. In such soils Mn toxicity is a major problems for crop production. Management of pH is very important in control of Mn toxicity. Mn toxicity symptoms include yellowing of older leaves, which darken into small, brown spots. Symptoms many times resemble typical bacterial leaf spots.

The key to optimize nutrient management lies in overcoming physiographic constraints to minimize loss of surface soil by erosion, adding adequate FYM augmentation by green manuring, judicious liming to avoid over-liming hazard, appropriate timing, split application and placement of fertilizers and use of foliar application of micronutrients as needed. Until site specific nutrient management is a reality, it will be practical to fortify fertilizer complexes enriched with Zn and B for successful production of vegetable crops.

Soil Health Improvement Through Crop Residue

India produces about 500 million tons (Mt) of crop residues annually (MNRE 2009). Processing of agricultural produce through milling and packaging also produces substantial amount of residues. Crop residues are natural resources with tremendous value to farmers. These residues are used as animal feed, composting, thatching for rural homes and fuel for domestic and industrial use. About 25% of nitrogen, 25% phosphorus, 50% of sulphur and 75% of potassium uptake by cereal crops are retained in residues, making them valuable sources of nutrients. However, a large portion of the residues, about 140 Mt, is burned in field primarily to clear the field from straw and stubble after the harvest of the preceding crop. The problem is severe in irrigated agriculture, particularly in the mechanized rice-wheat system. The main reasons for burning crop residues in field include unavailability of labour, high cost in removing the residues and use of combines in rice-wheat cropping system especially in the Indo-Gangetic plains (IGP). Primary crop types whose residues are typically burned include rice, wheat, cotton, maize, millet, sugarcane, jute, rapeseed-mustard and groundnut.

Burning of crop residues leads to 1) release of soot particles and smoke causing human health problems; 2) emission of greenhouse gases (GHGs) such as carbon dioxide, methane and nitrous oxide causing global warming; 3) loss of plant nutrients such as N, P, K and S; 4) adverse impacts on soil properties and 5) wastage of valuable C and energy rich residues. There are several options which can be practiced such as composting, generation of energy, production of biofuel and recycling in soil to manage the residues in a productive manner. Conservation agriculture (CA) offers a good promise in using these residues for improving soil health, increasing productivity, reducing pollution and enhancing sustainability and resilience of agriculture. The resource conserving technologies (RCTs) involving no- or minimum-tillage, direct seeding, bed planting and crop diversification with innovations in residue management are possible alternatives to the conventional energy and input intensive agriculture.

Crop residues production

Among different crops, cereals generate 352 Mt residue followed by fibres (66 Mt), oilseed (29 Mt), pulses (13 Mt) & sugarcane (12 Mt). The cereal crops (rice, wheat, maize, millets) contribute 70% while rice crop alone contributes 34% of crop residues. Wheat ranks second with 22% of residues whereas fibre crops contribute 13% of residues generated from all crops. Among fibres, cotton generates maximum (53 Mt) with 11% of crop residues. Coconut ranks second among fibre crops with 12 Mt of residue generation. Sugarcane residues comprising tops and leaves generates 12 Mt i.e., 2% of crop residues in India.

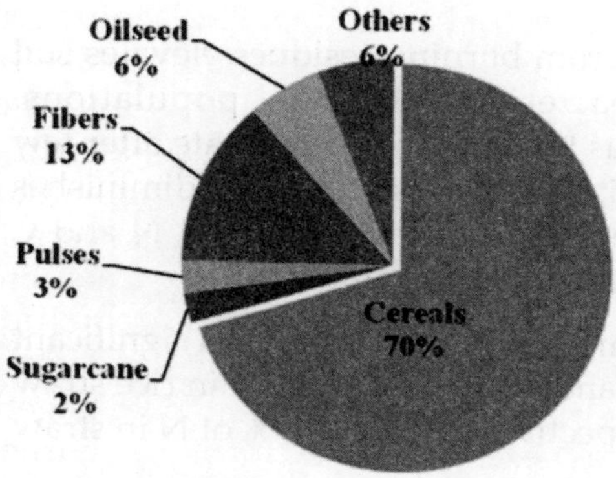

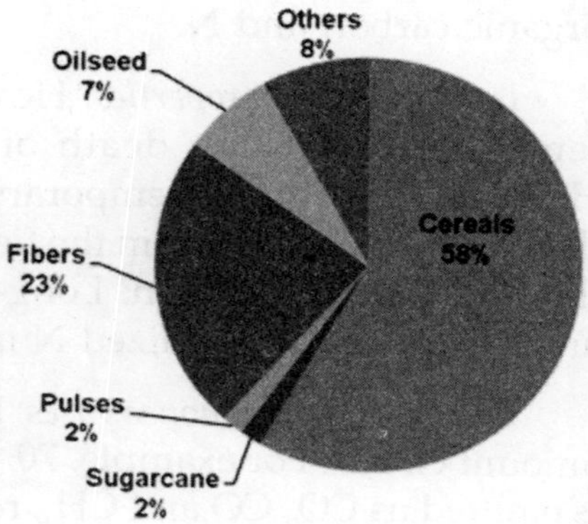

Fig 15.1 Contribution of various crops in residue generation in India (Calculated from MNRE report 2009).

The surplus residues i.e., total residues generated less residues used for various purposes, are typically burned in the field or used to meet household energy needs by farmers. Estimated total crop residue surplus in India is 84-141 Mt yr^{-1} where cereals and fibre crops contribute 58% and 23%, respectively. Remaining 19% is from sugarcane, pulses, oilseeds and other crops.

Causes of burning crop residues in field

Increased mechanization, particularly use of combine, declining number of livestock, long period required for composting and no economically viable alternate use of residues are some of the reasons for residues being burnt in field. Clearing of fields, fertility enhancement, pest and pasture management. Burning traditionally provides a fast way to clear the agricultural field of residual biomass and facilitating further land preparation and planting. It also provides a fast way of controlling weeds, insects and diseases, both by eliminating them directly or by altering their natural habitat. The time gap between rice harvesting and wheat sowing in northwest India is 15-20 days. In this short duration farmers prefer burning the rice stalk in the field instead of harvesting it for fodder.

Adverse consequences of residue burning in field

Loss of nutrient : In addition to loss of entire amount of C, 80% of N, 25% of P, 50% of S and 20% of K present in straw is lost due to burning, it also pollutes the atmosphere. If the crop residues are incorporated or retained, the soil will be enriched, particularly with organic carbon and N.

Impact on soil properties: Heat from burning residues elevates soil temperature causing death of bacterial and fungal populations. However, the death is temporary as the microbes regenerate after few days. Repeated burning in the field, however, permanently diminishes the microbial population. Long-term burning reduces total N and C and potentially mineralized N in the 0-15 cm soil layer.

Emission of greenhouse gases: Burning of residues emits a significant amount GHGs. For example, 70, 7 and 0.66% of C present in rice straw is emitted as CO_2, CO and CH_4, respectively, while 2.09% of N in straw is emitted as N_2O upon burning.

Beneficial effects of residues on soil health and crop yield

Incorporation of crop residues in soil or retention on surface has several positive influences on physical, chemical and biological properties of soil. It increases hydraulic conductivity and reduce bulk density of soil by modifying soil structure and aggregate stability. Mulching with plant residues raises the minimum soil temperature in winter due to reduction in upward heat flux from soil and decreases soil temperature during summer due to shading effect. Retention of crop residues on soil surface slows runoff by acting as tiny dams, reduces surface crust formation and enhances infiltration. The channels (macropores) created by earthworms and old plant roots, when left intact with no-till, improve infiltration to help reduce or eliminate runoff. Combined with reduced water evaporation from the top few inches of soil and with improved soil characteristics, higher level of soil moisture can contribute to higher crop yield in many cropping and climatic situations.

Residues act as reservoir for plant nutrients, prevent leaching of nutrients, increase cation exchange capacity (CEC), provide congenial environment for biological N fixation, increase microbial biomass and enhance activities of enzymes such as dehydrogenase and alkaline phosphatase. Increased microbial biomass can enhance nutrients availability in soil as well as act as sink and source of plant nutrients. Leaving substantial amounts of crop residues evenly distributed over the soil surface reduces wind and water erosion, increases water infiltration and moisture retention, and reduces surface sediment and water runoff.

The crop residues play an important role in amelioration of soil acidity through the release of hydroxyls especially during the decomposition of residues with higher C:N ratios, and soil alkalinity through application of residues from lower C:N ratio crops including legumes: oilseeds and pulses. The role of crop residues on carbon sequestration in the soil would be an added advantage in relation to climate change effects management.

Yield response with residue management varies with soil characteristics, climate, cropping patterns, and level of management skills. Greater yields with residue application results from increased infiltration and improved soil properties, increased soil organic matter and earthworm activity and improved soil structure in 4-7 years from when the system is established.

Crop residues management for conservation agriculture

Permanent crop cover with recycling of crop residues is a prerequisite and integral part of CA, which is advocated as alternative to the conventional production system for improving productivity and sustainability. Recent estimates revealed that CA based resource conserving technologies (RCTs) that include laser assisted precision land levelling, zero/reduced tillage, direct drilling into the residues, direct seeded rice, unpuddled mechanical transplanted rice, raised bed planting and diversification/intensification are being practiced over nearly 3.9 Mha of South Asia. The RCTs with innovations in residue management avoid straw burning, improve soil organic C, are input efficient and have potential to reduce GHG emissions.

New variants of zero-till seed-cum-fertilizer drill/planters such as Happy Seeder Turbo Seeder and rotary-disc drill have been developed for direct drilling of seeds in presence of surface residue (loose and anchored up to 10 t ha^{-1}). These machines are very useful for managing crop residues for conserving moisture and nutrients and controlling weeds as well as moderating soil temperature.

Alternative uses of crop residues

There are several options which can be practiced to manage residues in productive manner. Besides use as cattle feed, large amount of residues can be used for preparation of compost, generation of energy and production of biofuel and mushroom cultivation.

Composting of residues for manure: The residues can be composted by using it as animal bedding and then heaping in dung pit. Each kg of straw absorbs about 2-3 kg of urine from the animal shed. It can also be composted by alternative methods on the farm itself. The residues of rice from one hectare give about 3.2 tons of manure as rich in nutrients as farmyard manure (FYM).

Energy from crop residues: Biomass can be efficiently utilized as a source of energy & is of interest worldwide because of its environmental advantages. During recent years, there has been an increase in the usage of crop residue for energy production and as substitute for fossil fuels.

Ethanol from crop residues: The conversion of ligno-cellulosic biomass into bio-based alcohol production is of immense importance and is a researchable issue as ethanol can be either blended with gasoline as a fuel extender and octane-enhancing agent or used as a neat fuel in internal combustion engines. The theoretical estimates of ethanol production from different feedstock (corn grain, rice straw, wheat straw, bagasse and saw dust) varies from 382 to 471 L t^{-1} of dry matter.

Biomethanation: Biomass such as rice straw can be converted to biogas, a mixture of carbon dioxide and methane and used as fuel. It is reported that biogas of 300 m^3/tonof dry rice straw can be obtained. The process yields good quality of gas with 55-60% of methane and the spent slurry can be used as manure. This process promises a method to utilize crop residues in a non-destructive way to extract high quality fuel gas and produce manure to be recycled in soil.

Biochar: Biochar is high carbon material produced from the slow pyrolysis (heating in the absence of oxygen) of biomass. It has got advantages in terms of its efficiency as an energy source, its use as a fertilizer when mixed with soil, its ability to stabilize as well as reduce emissions of harmful gases in the atmosphere. It can potentially play a major role in the long-term storage of carbon. Biochar increases the fertility, water retention capability of the soil as well as increasing the rate of mineral delivery to roots of the plants.

□□□

16

Technology Up-scaling and Participatory Approach

An array of technologies have been developed and validated for enhancing on-farm water productivity. Technologies ranging from input selection and management to marketing of a product have been standardized. The major issue that undermines the potential of these technologies is the adoption by the farmers. The traditional system of extension which was so effective in the early years of green revolution was based on the development of recommendations and packages and then demonstrating it to the farmers. Over the years there has been significant refinement in extension approaches to demonstrate the technologies. These approaches are critical to adoption of technologies related to improving water productivity. This chapter discusses the approaches to upscale the technology. Also, the aspects of orchard management through cluster approach of people's participation is discussed.

Upscaling is a form of technology transfer process through which existing technology is translocated /disseminated from its source or intermediary to the final user. When we talk about upscaling process it means we want to upscale or increase the level of adoption of a technology. In other word it is exposing technology to a wider group of recipients*i.e* beyond the experimental/technology generation group. Sometimes outscaling and upscaling are used synonymously. Generally outscaling is used for dissemination of technology form one system to another or from one locality to other locality for larger coverage in terms of area or number of people. Technology upscaling is achieved through:

- Replication Dissemination, spread, or

- Adaptation of techniques,
- Ideas and approaches.

Challenges in technology upscaling

Many technologies are confined to limited area or adopted by limited stakeholders due to one or another reasons; some of the challenges to upscale agricultural technologies in hill and plateau regions are:

- Weak linkages among stakeholders engaged in upscaling
- Lack of active participation by actual farmers
- Inadequate communication plans for promotion of uptake and scaling up;
- Inadequate evaluation for uptake and use of desired knowledge;
- Inadequate fund and human resources for promotion of uptake and scaling up;
- Inadequate capacity in promotion of uptake and scaling up;
- Failure to link reward and incentive systems to impact;
- Insufficient end-user involvement;
- Ineffectiveness in the extension systems and the technology dissemination processes;
- Poor identification of media and channels for communicating with different stakeholders.

There should be proper strategy and plan to combat those challenges of upscaling of agricultural technologies in hill and plateau regions and accelerate dissemination process for wider adoption of technologies suited to the area by the stakeholders. Following strategies may be followed for the hill area:

- Determination of existing resources
- Prioritization of needs
- Identification of Knowledge gap
- Creating awareness;

- Capacity building of field personnel/FIG/Change agent
- Identification of local stakeholders engaged in agricultural development
- Convergence with ongoing schemes
- Linkage for creditAssurance of seed & planting materials
- Marketing and storage facilities
- Develop plan for upscaling alongwith target groups
- Discuss plan for upscaling with stakeholders
- Engaged actors/agencies responsible/helpful in upscaling
- Follow up/monitoring of upscaling
- Popularize success cases as model for upscaling.

Outscaling through Innovation Platforms

An innovation platform is a space for learning and change. It is a group of individuals (who often represent organizations) with different backgrounds and interests: farmers, traders, food processors, researchers, government officials etc. (*ILRI Practice Brief #1*).The members come together to diagnose problems, identify opportunities and find ways to achieve their goals. They may design and implement activities as a platform, or coordinate activities by individual members dentification of stakeholders engaged in agricultural development in the area can help in upscaling of technologies in the area.

Table 16.1 Actors involved in promoting agriculture in a district

Sector	Organization names	
Public	Government Organization (Research &Extension) Department of Agriculture Department of Horticulture ATMA Birsa Agricultural University KVK Panchayat: rural infrastructure/ development District Rural Dev. Agency (solar pumpsets) Water resource department.	Government Projects Department of Agriculture responsible for RKVY,NFSM & others state run schemes Department of Horticultural responsible for National Horticultural & Micro-irrigation Missions NABARD Women Empowerment & Livelihood Program

Civil	NGOs Ramakrishna Mission with its 50 plus staff Other potential NGOs engaged in community development	Community & Farmer Organizations SHGs & Village Organizations Women SHGs
Private	Input suppliers/output marketers Retailers at district, block & village level, Department of Agriculture registered wholesalers (100) Jain Irrigation and other Agricultural Machine dealers/ manufacturers.	Credit Agencies NABARD Farmer Clubs (300) & Joint Liability Groups(500) Commercial banks (20) with 200 branches PACS (where functional) Micro-credit agencies.

Approaches for technology upscaling

Individual to individual: Personal call, farmer to farmer communication, Phone- in–live programme is useful.

Individual to group: Personal calls, farmer to farmers' communication, E-Mail, Mobile messages, Exposure visits are useful.

Group to individual: Farmers- Scientist interaction

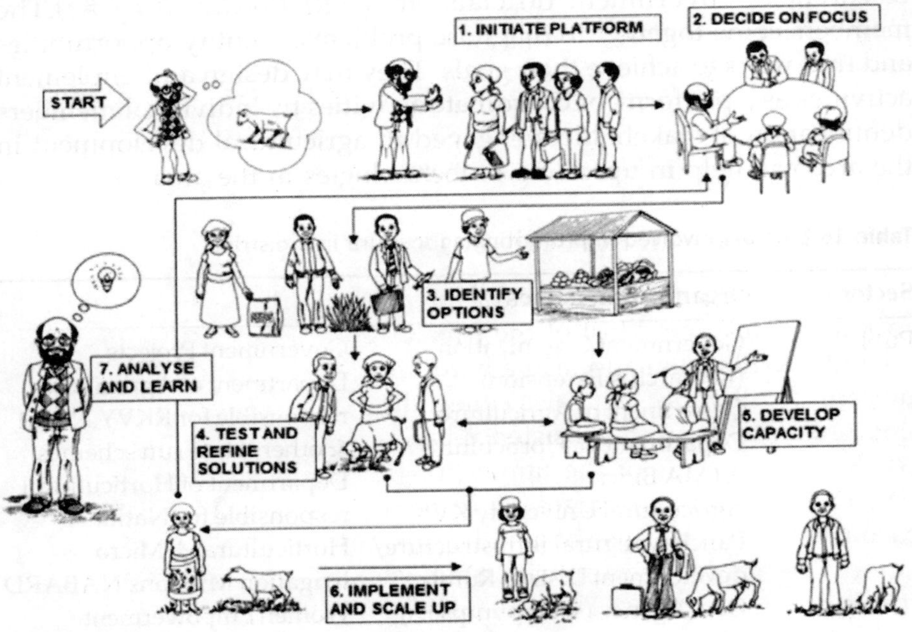

Fig 16.1 Stages of outscaling(*Source: ILRI, 2013)*

Group to group: Farmers field school, Exhibitions, Mobile messages, Mass media especially radio etc. are useful.

Upscaling strategies for hill and plateau regions should be resource based and targeted for different categories of farmers involving women and rural youth. Different stakeholder (farmers, SHGs, NGOs etc) and agencies should be involved for the upscaling.

Table 16.2 Upscaling strategies for different categories of farmers in a hill area for livelihood improvement

Category	Current livelihood	Possible interventions/ strategy
Landless	Wage labour, Lease in land	Targeted wage employment programmes through construction of water harvesting structures, reshaping of ponds, terracing, engaging in vegetable cultivation, preparation of planting materials, mushroom cultivation.
Small & medium Farmers	Wage labour	Skill up gradation, Resource conservation through watershed approach, vegetable nursery raising, hybrid seed production, diversification, Integrated farming etc.
Well to do farmers	Traditional farming	Integrated farming, Resource recycling, precision agriculture, high density orchards establishment, protected cultivation with marketing arrangement etc.

Cluster Approach for Horticulture Development

Promoting horticulture in rural and tribal dominated belts of India is a major challenge. Most of the tribalpopulation is poor, backward, generally uneducated and lead a hard life. They practice both shifting and settled cultivation, and also collect the minor forest produce items. Food grain yields from subsistence-based farming are not enough to meet the food needs of the family for the whole year. The periodic cycle of food grain shortages represents the "lean period" or "lean season" for them. Because of a higher incidence of poverty, they failto acquire adequate calories, nutrition and vitamins to keep them immune from various diseases. Poverty further diminishes their earning capability.

The economy of tribal people is based on output from shifting cultivation and forest including some other alternative livelihood options. Agriculture and forest, being important livelihood options, a significant proportion is engaged in such practices. Due to high slopes and lack of conservation measures, soil erosion is very high in such high lands, as such; the productivity of land and soil fertility has declined over the years. Reducing ecology and livelihood inter-relationship is spiraling down gradually with continued decrease in carrying capacity of the available natural resources in the area.

Ecological degradation, erratic rainfall and high risk of drought have resulted in food insecurity and increasing out-migration. Land alienation, forest legislation, mining operations, irrigation projects, wildlife and sanctuaries etc have also affected the tribal livelihood. More so, they reside in environmentally adverse and physically less accessible areas with scattered human settlements, they are more vulnerable to various types of endemic, contagious and seasonal diseases due to mass poverty, ignorance and absence of proper health care infrastructure and services.

Various horticultural crops including mango is being promoted in remote tribal areas where traditional cultivation is tough and less profitable.

The agriculture land in each village can be classified into:

- Hillocks and Upland (*Dongar*): Hillocks are generally fallow. Shifting cultivation of niger, ragi, cowpea, fallow, pulses, oil seeds, cashew are main components of such land. Diversity loss, soil erosion and low productivity are the major problems.
- Homestead land (*Bari*): Baris are used for kitchen garden crops. Mixed cropping in Kharif (maize, vegetables) followed by short duration crops using residual moisture. Low production and small homestead are the problems.
- Medium up land (*Bhata*): Such land is used for short duration paddy (local), millets, niger, tuar and cashew. Crop failure grazing, moisture stress and low production are the major problems.
- Medium lowland (*DhepaKyari*) : Such lands are used for medium duration paddy and many other crops. Seasonal irrigation, low production, grazing and poor fertility are the problems.

- Valley land (*Kyari*): Such lands are irrigated and remain occupied round the year. These mainly belong to well off households. Relatively long duration paddy is cultivated but productivity is low.

Paddy, Arhar (Kandul), Mandia, ragi, mung, biri, kulthi,Kating, cowpea, Rashi (til), mustard, niger, groundnut, potato and maize are the major crops grown in the area. Mango, banana, vegetables, turmeric, ginger, oranges land races of cattle and poultry are also common. It was recorded that rice is grown by only a few households having valley land; millets locally known as *sawan*, ragietc, is grown by majority of farming families. Nearly 42.91 percent families grow oil seeds (especially niger and mustard).

With a vision to make horticulture as a primary source of livelihood for tribals, attempts were made in mission mode to develop diversified fruit belt. It was demonstrated that mango can be a source of income for tribalsfrom the land which is otherwise fallow. Fruit fly incidence, poor confidence of tribals in mango, shortage of planting material, lack of knowledge on cultivation technology on sloppy land, quality maintenance and lack of marketing channel were the major impediments in the area.

Need based technological interventions were made from field to fork on cluster basis. The efforts have been transformed in to outcomes as mango has become as a priority crop for tribals in the area.

First Step – Community mobilization

The approach of the work included system analysis of tribals in villages for identification of prevalence of faming system, its strength and scope of improvement with special emphasis in mango based horticulture. The programme also included the laboratory related works at research station such as, standardization of technologies in relation to tribal area and then modification and execution of elsewhere prevailing technologies in the target area. Social mobilization under leadership of tribal society and progressive farmers, Group (cluster) approach on village basis for tribal participation, pest and disease management, planting methodology, quality production and marketing support were the key issues during the work period. The schedule of activities as performed was as under:

- Establishing the role of horticulture in the livelihood of tribals.
- Identification of critical food shortage period for scope of intervention.
- Social mobilization for tribal sensitization.
- Technological standardization, refinement and field execution for pest management and quality mango production.
- Technological support for area expansion under mango plantation.
- Soil and moisture conservation and plant health management under rainfed sloppy land.
- Quality maintenance through Hot Water Treatment (HWT) and packaging of produce.
- Development of channel for distant area marketing.

Linkage with stakeholders

1. Tribal groups of villages of Kashipur.
2. Horticulture & Agricultural Related Panchayats Association (HARPAL), Kashipur, Rayagada (Local logistics & motivator).
3. Central Horticultural Experiment Station (IIHR), Aiginia, Bhubaneswar (Technology provider).
4. Nation Horticulture Board, Nayapalli, Bhubaneswar, Odisha (Funding for some plantations).
5. Indian Agricultural Research Institute, Pusa, New Delhi (Post harvest management works).
6. National Research Centre on Pomegranate, Solapur, Maharashtra (For pomegranate technology).
7. Directorate of Horticulture, Bhubaneswar, Odisha.

The work was carried out in Kashipur block of Rayagada district of Odisha state. A total of 9 Gram Panchayats, scattered in 26 hamlets were covered. Out of 529 farmers covered under the programmes, 384 were tribals making a percentage of 72.58 tribal families.

The tribal led society named as HARPAL helped as social mobilizer, produce collector, packer, marketer and field executor whereas Central

Horticultural Experiment Station –a regional station of Indian Institute of Horticultural Research under Indian Council of Agricultural Research served as technology developer, its provider and demonstrator. The progressive farmers of the society also participated in field evaluation of technologies. The assistance of Department of Horticulture Orissa was received as a monitor of technological execution in the field and Mahalaxmi fruit store, Bhubaneswar helped as consultant in getting the feedback of retailers and consumers preference pertaining to mango packing, retailability, quality and fruit perishability. Lead farmers served as first risk bearer and tribal farmers groups of different villages under SHG mode participated in the work. In between some intervention from APEDA was also received.

Second Step – Problem Identification

Constraints were perceived as the factors which are putting the negative effects on the farming systems and elimination of such factors would strengthen the farming system. These were identified as lack of land title for the land under shifting cultivation in the name of tribal households, depletion of soil fertility under shifting cultivation, poor know how and technical knowledge on mango, low level promotion of perennial horticulture crops, non-availability of good quality planting materials, heavy incidence of fruit fly, incidence of anthracnose due to late ripening, poor canopy management and absence of bulk mango marketing channel due to distant location of production area. It was seen that the forest areas are degenerated due to shifting cultivation, hillocks and foot hills are naked and abandoned as non-cultivable land. The foot hills and hillocks have gravels as most of the top soil is washed away rendering it unfit for field crop cultivation. These high lands are well suited only for tree based perennial horticultural crops. Attempts were made to fill this gap over the period.

Third Step - Capacity Building

Training on various aspects of mango cultivation, mango day, farmers gathering in tribal mango growers' field on recurring basis, group meeting in village panchayats, personal persuasion and confidence building visits were some of the methods for tribal mobilization. The livelihood decision is basically based on women decision in tribal households; hence tribal women sensitization for mango cultivation was also done in the villages. The male workers had no say in land belongingness/topography vis a vis mango planting.

Impact of training was recognized from the fact that the nurseries raising started by many farmers. In addition, progressive farmers, on the demand of tribals started nursery where all the trainings were imparted and lakhs of grafts were prepared to distribute free of cost among the tribals. A team of 30 local tribal boys was trained in mango rootstock development and grafting. As a result, large number of Amrapalli grafts were made during the period which was given to tribals free of cost.

Fourth Step – Adoption and active participation

The motivation and demonstration to famers in previous years, led to the increase in the mango area. Camps on mango cultivation were organized in the villageswith the help of NHM in 2006. Again meetings of tribals of 10 villages were convened in fruiting mango orchards which motivated them for mango planting *en mass*. Planting materials were available locally and the planting technology was demonstrated in the field itself in all the villages. Thus, the planting in large area was done as per decision taken by the tribal groups/village leaders.

Some of the tribals wanted to plant mango on land having acute slope. As the area has undulated topography and poor or no source of irrigation, attempts were made to develop *insitu* grafted mango orchards on such lands with all the soil and moisture conservation techniques.Techniques used were making platforms for individual plant, deep and wide pit digging for each plant, removal of gravels while digging the pit and use of proper pit filling mixture, *in-situ* raising of root stock (3 nos./pit) and grafting with elite variety, developing *in-situ* water conservation structures near plant basin, grass belt conservation and Sod culture, mulching with available grasses in the plant basin and fire protection by cutting the strip of grasses around the hillocks. It is needless to mention that such soil work require a lot of manpower. Interestingly, all these works were done by tribals only, though no money was paid to them for such works. This indicated a high level of motivation among them. The structures on slope need recurring and annual maintenance which they are doing. The growth of plants under such situation is much better than non-managed plants.

Fifth Step- Post adoption Interventions

Zero orchard tillage to reduce erosion, planting under high density of 5x5 m distance in 1x1x1 m pit size with standard practice of pit filling, nutrient supply and plant canopy architecture techniques have been adopted. As the orchards are located in sloppy land, foot hills and hillocks, *in situ* rain water harvesting technique i.e. half-moon basin structure with front trenching have been adopted which helps holding in 5-6 buckets of rain water. Orchard management practices like mulching with grasses, providing live fencing and standard canopy management have been applied to plants during this period.

Green fruits were harvested at appropriate stage of maturity. After harvesting the fruits were kept to discharge sap and then kept in plastic crates. These fruits were subjected to hot water treatment plant, constructed in the area itself. The hot water treatment plant has been manufactured locally, using 2000 liters capacity zinc tank fitted with heating rods, thermostat and circulating fan to mix the water to assure uniformity of temperature within the tank. Using pulley and chain, the mango crates were submerged in hot water tank 10 inches below the water surface. The ratio of volume of water to fruit stock was kept large enough to prevent a significant drop in temperature when the mango stock is added. Fruits were treated at 46^{0}C for 60 min and were allowed to cool and dry before grading and packing. In one shift, 0.55 ton mango is treated and such 18-20 shifts were managed in a day. In India, most of hot water treatment plants are meant for export of mango, however, this is the only HWT plant which provides mango for domestic consumption.

After cooling and drying, fruits were packed in corrugated card board boxes of 5 kg capacity. Paddy straw was used as cushioning material for the packaging of fruits. Initially packing was done in beer card board packs; however, HWT and transportation in these packets did not go easily. Heavy loss was sustained due to heat injury and pressure damage. Then the hot water treatment system and packing was standardized to avoid the loss and mango packing started in 4 kg cardboard packs as suggested by wholeseller. Mangoes are loaded in trucks (6 ton per trip) for transportation and sale up to more than 500 km in urban areas. Views of fruit stockiest and retailers were collected pertaining to packing, perishability and retailability of the produce. The consumers' preference was also collected in an informal way. Based on market reports, the fruits are now transported in plastic crates of

20 kg and then repacking is done in urban areas for making family size pack.

Principal milestones reached

- Availability of QPM of mango locally.
- In situ mango orchard establishment technology on sloppy land.
- In situ water harvesting techniques and soil conservation practices on sloppy land under rainfed condition.
- Area expansion of mango.
- Perfected fruit fly management technology.
- Hot Water Treatment technology of mango.
- Quality produce with extended shelf life fetching more price than others.
- Market channel for hot water treated mango.
- Social awareness and confidence of tribals towards in economic aspects of mango and value of waste land.

Impact

Capacity building in production and use of own QPM in tribal villages: The nurseries were developed in local villages. In addition, lead farmers, on the demand of tribals started nursery where all the trainings were imparted and lakhs of grafts were made to distribute free of cost among the tribals. The number of plants planted out of own produced grafts are presented in Table 4.

Mango plantation area expansion: The total area brought under mango plantation during the work period is presented in Table-5. It is seen that a total of 1214 acres of land was brought under mango plantation of which, 843 acres belong to tribals which comes to 69.60 per cent of total planted area.

Canopy management: As all the plantations are under high density, all the farmers who planted mango have been trained to make initial frame of mango so that problem of overlapping and overcrowding does not appear in days to come. Pruning in grown up plants and severe pruning in overcrowded orchards have also been done.

***In situ* orchard on sloppy land under rainfed condition:** As on trial basis an orchard of 15 acres was developed *in situ* on gentle hillock having more than 30 percent slope. As on today, this orchard is fruiting well. Looking to the success, some of the hillocks having gentle slope, have been planted with mango under *in situ* condition.

***In situ* water harvesting in fields and moisture conservation:** Much of the plantations, covering area more than 900 acres have been supplemented with front trench and half moon soil and water conserving structures against the slope. Care has been taken that each plant has its own water catchment area. Likewise, trenches have been made in the orchards across the slope to reduce the soil erosion. It is seen that the growth of such plants is much better than others.

IPM of fruit flies and other pest management practices: The technology standardization process started from 2002. During the process of technology standardization, many farmers came to know the process and effectiveness of the technology and started its use. MAT and BAT together can give 95 per cent protection and there is no need of use of pesticide in mango orchard for fruit fly control. Pesticide: produce ratio in MAT and BAT is 19.02 kg fruit per ml of insecticide whereas only 0.38 in cover spray.

Inter-cropping of tubers: Not in large area but in some fields, tapioca and sweet potato have been introduced as intercrops on the bund of trenches.

Hot water treatment of mango: The HWT was essential to extend market life even if mango was pre-treated in the orchards. Normally ripening of mango starts by 25th June in the said area but rains commence from 15th June. Being hot, humid and high rainfall area (1500mm), the incidence of anthracnose is high. Hot Water Treatment eradicated the disease inoculums from fruit surface and brought about reduced enzymatic activity therefore, decay process in mango was reduced. Further, the possibility accidental fruit fly larval presence in ripe fruits was also ruled out, hence longer market life.There was heavy reduction in incidence of fruit fly and diseases in mature mango fruits due HWT as such treated fruits were appreciated by consumers.As reported by retailers, marketability of un-treated fruits reduced much early than treated ones, therefore, poor sale ability of untreated fruits.

Packaging and marketing of produce: Initially packing was done in beer card board packs; however, HWT and transportation in these packets did not go easily. Heavy loss was sustained due to heat injury and pressure damage. Then the hot water treatment system and packing was standardized to avoid the loss and mango packing started in 4 kg card board packs as suggested by whole seller and retailers and finally in plastic crates of 20 kg.

Nutritional improvement: The direct impact of nutritional improvement has not been assessed, however it is understood that cash inflow during the food shortage period have increased the purchasing power and thereby adequate food and nutrition. In addition to sale, every farm family also enjoyed the mango which directly influences the beta carotene intake in addition to other nutrients

Trainings organized: Various trainings, scientist visits and farmers day were organized during the work period. Most of the trainings were conducted as method demonstration at field itself. The training schedule included method demonstration in fields and result demonstration. Frequent visits to field and on spot training were found to be learning and motivatethetribals.

Sustainability

- No material is brought from outside except fertilizer and pesticides and every activity is performed in natural environment. There is no coercive technology, rather ecosystem restoring and livelihood securing technologies are in practices, hence are sustainable.
- Local groups of tribals has been trained in various techniques such as fruit fly management, hopper and powdery mildew management, nursery technology, planting technology, soil and water conservation technology and canopy management, hence for most of these technologies, they need not to depend on external agencies.
- As the potential of the mango cultivation has been realized by tribals from their neighbours, they do not need sensitization. They are now doing it by their own.
- HWT plant is running and treated mango is sold out every year, however 2-3 more HWT plants are needed in the area.
- Area expansion under mango is going on.

- Mango from this area comes when Andhra mango is receding and mango from Northern states is not available in the market, hence being located at advantageous position, area has tremendous potential which has been realized by the tribals, hence the disseminated technology will remain sustainable.
- There is a demand to develop facilities for ripening and packaging infrastructure along with pack standardization for different varieties which indicates the involvement of tribals in the said work.
- There is a permanent channel between tribal group, lead farmer, research institution and market; hence, the process will go ahead in days to come also.

New initiatives / proposals and future plan

- Establishment of Training Centre at Kashipur to be funded under RKVY by Government of Odisha.
- Establishment of PHM centre (Desapping, washing, HWT, grading and packing (To be established by IARI, New Delhi).
- Developing irrigation facility for 6 villages for horticulture (Surface water channelizing, project submitted under RKVY by DDH, Rayagada).
- Establishment of input supply centre (To be established near training centre).
- Testing of fruit crops such as pomegranate, Nagpur orange, kinnow, litchi, coloured grapes and pine apple for their feasibility and commercial potential.
- Mango from this area comes when Andhra mango is receding and mango from Northern states is not available in the market, hence being located at advantageous position area can be developed as a mango belt.
- Plantations are being done on sloppy land having 10-50% slope, it will need good management of soil and water.
- Micro-nutrient management (as foliar spray) schedule is needed as most of the soils are poor and have disadvantageous location.
- All fields are having High density plantation which needs perfect canopy management strategies.

- Many areas are prone to hoppers, stone weevil, fruit borer and fruit flies, hence proper techniques are needed to contain these pests.
- Disease awareness and especially anthracnose and powdery mildew is needed.
- Hot water treatment Plant.
- Empowering the grower for self-marketing rather than contract marketing.
- Pack standardization for different varieties and facilities for packaging.
- Developing new marketing channels outside the state especially in south India.
- Developing processing industry when bulk mango starts to come.

Horticultural development has been a priority in the area but without motivation and result demonstration, hence, most of the land is still vacant despite massive efforts by Govt. Block has a vast area of Dongar land (natural resource), which has gone out of use due to soil erosion and no field crops are grown in this land, therefore horticulture is the only option for these lands. There is ample scope of fruit cultivation in tribal belts, especially where shifting cultivation, lack of irrigation and soil erosion are prevalent. Jobless man power (human resource) is available in the area where livelihood options are shrinking. Many tribals are willing for fruit crops but they have technological, financial and marketing constraints. The technology (skill based human resource) can be developed, tested and made available in the area. The technological interventions can play great role in making not only mango but other horticultural crops a profitable enterprise. The result demonstrations and social mobilization can develop confidence among tribals in fruit crop cultivation. The market channel plays a vital role in adoption of horticulture based entrerpises. Ignorance or stigma is not a cause of poor adoption of fruit crop cultivation among the tribals of Orissa but lack of result oriented demonstration, motivation and confidence.

□□□

References

Aarti K, Dixit S, Greco R, Asaph A, Trijatmiko KR., Martinez NM, Krishnan A, Nataraja KN, Udayakumar M, Pereira A. 2007. Improvement of water use efficiency in rice by expression of HARDY, an Arabidopsis drought and salt tolerance gene. *Proceedings of National Academy of Sciences* 104(39): 15270-15275.

Agarwal RG. 2007. Food Security and role of agrochemicals. *Crop Care* 33 (2):3-8.

Aggarwal PK, Joshi PK, Ingram JSI and Gupta R K. 2004. Adapting food systems of the Indo-Gangetic plains to global environmental change: Key information needs to improve policy formulation. *Environ. Sci. Policy*. 7: 487-498.

Akey WC, Jurik TW, Dekker J. 1990. Competition foe light between velvetleaf (*Abutilon theophrasti*) and soybean (*Glycine max.*). *Weed Research* 30: 403-411.

Almezori HAM. 1996. Studies on the allelopathic potential of *Zea mays* L. Ph.D. thesis Baghdad University, Iraq.

Alsaadawi IS, Mahdi AS, Bapeer UHK. 1993. Separation of phytotoxins from *Sorghum bicolor* L. residues, and study of its persistence in soil. 1st Sci. Conf. field crops Res. Baghdad, Iraq.

Annon. 1998. Production, processing and marketing worldwide. Poultry International Watt Publication Co 18 Chapel Street, Hampshire, England.

AR. 2012. *Annual Report 2012 - 2013*. Department of Animal Husbandry, Dairying and Fisheries, Ministry of Agriculture, Government of India, Krishi Bhawan, New Delhi.

Asaph A, Dixit S, Jetter R, Thoenes E, Arkel GV, Pereira A. 2004. The SHINE Clade of AP2 domain transcription factors activates wax biosynthesis, alters cuticle properties, and confers drought tolerance when overexpressed in Arabdiopsis. *The Plant Cell* 16(9): 2463-2480.

Avesta, M, Ali Hassan, AMMezori. 2008. The allelopathic potential of some crops and vegetables in mixed farming in Dohuk governorate I. water extract. *J. Dohuk Univ.* 11-1.

Baig, MJ, Singh S, Rai SK, Singh JB, Misra AK, Biradar NB, Verma OPS, Kumar A. 2007. Water productivity in forage crops and livestock: Prospects and opportunity for improvement. Indian Grassland and Fodder Research Institute, Jhansi, India.

Barberi, P. 2001. Weed management in organic agriculture: are we addressing the right issue? *Weed Research* 42: 177-193.

Behera, UK, Bhan S. 2014. Conservation agriculture in India – Problems, prospects and policy issues. *International Soil and Water Conservation Research.* 2:4 (1-12).

Benbi, DK, Chand, M. 2007. Quantifying the effect of soil organic matter on indigenous soil N supply and wheat productivity in semiarid sub-tropical India. *Nutrient Cycling in Agroecosystems,* 79: 103-112.

Betts, RA, Boucher, Collins, M. Cox PM, Falloon, PD, Gedney, N, Hemming, DL, Huntingford C, Jones CD, Sexton DMH, Webb MJ. 2007. Projected increase in continental runoff due to plant responses to increasing carbon dioxide. *Nature* 448:1037–1041, doi:10.1038/nature06045.

Bhatnagar, PS, Joshi, OP. 1998. Intercropping soyabean for higher returns. Ministry of Agriculture, GOI, Krishi Vistar Bhawan, New Delhi.

Bilal, AK, and Wani, AA. 2011. Carbon Sequestration Potential of Biomass under different agroforestry land use systems in Poanta area of Himachal Pradesh. Paper presented at Indian Forestry Congress 2011, New Delhi, India.

Blum, A. 2005. Drought resistance, water-use efficiency, and yield potential are they compatible, dissonant, or mutually exclusive? *Aus. J. Agri. Res.* 56: 1159–1168.

Blum, A. 2009. Effective use of water EUW and not water-use efficiency WUE is the target of crop yield improvement under drought stress. *Field Crops Research.* 1122–3: 119-123. http://dx.doi.org/10.1016/j.fcr.2009.03.009.

Boatman, C, Stoate, R, Gooch, CR, Carvalho, R, Borralho, G, de Snoo, PE. 1999. The environmental impacts of arable crop production in the European Union: practical options for improvement. A report prepared for Directorate-General XI of the European Commission (1999).

Bora, DKD, Saharia, Hussain, S. 1994. Biology of yellow stem borer *Scirpophaga incertulas* Wlk. Crop Res Hissar 8: 366-370.

Briggs, LJ, Shantz, HL. 1913. Th e water requirement of plants. I. Investigationsin the Great Plains in 1910 and 1911. USDA Bur. Plant Ind. Bull.284. USDA, Washington, DC.

Briggs, LJ,Shantz, HL. 1914. Relative water requirement of plants. J.Agric. Res. (Washington, DC) 3:1–64.

Buchanan, GA. 1976. Management of weed pests of cotton (*Gossypiumhirsutum)*. In:Proceedings of US-USSR Symposium on the integrated control of the arthroped, disease and weed pests of cotton, grain, sorghum and deciduous fruits, Lubbock T.X. pp. 168-84.

Chadhar, S K and Sharma, M C.1996. Survival and yield of four medicinal plant species grown under tree plantations of Bhataland. *VanikiSandesh*. 20(4): 3–5.

Chaturvedi, RK, Joshi, J, Jayaraman, M, Bala, G, Ravindranath, NH. 2012. Multi-model climate change projections for India under representative concentration pathways. *Curr. Sci.* 1037:791-802.

Chauhan, SK, Sharma, SC, Beri, V, Yadav, S, Gupta, N. 2010. Yield and carbon sequestration potential of wheat *Triticum aestivum* poplar *Populus deltoides* based agri-silvicultural system. *Indian Journal of Agricultural Sciences* 80: 129-35.

Chhokar, R S, Sharma, RK, Jat, GR, Pundir, AK, Gathala, MK. 2007. Effect of tillage and herbicides on weeds and productivity of wheat under rice–wheat growing system. *Crop Prot*. 26:1689–1696.

Churchil, RR, Narayanankutty, K. 2003. A report on AICRP on poultry. Kerala Agric. University, Mannuthy, Kerala, India.

CIFA, 2013. Vision 2050. Central Institute of Freshwater Aquaculture, Bhubaneswar.

CRIDA. 1993 & 1996. Annual Reports. Central Research Institute for Dryland Agriculture CRIDA, Hyderabad, India.

CRIDA. 2007. Perspective Plan, Vision 2025. Central Research Institute for Dryland Agriculture, Hyderabad, Andhra Pradesh.

Croissant, RL, Peterson, GA and Westfall, DG. 2014. Dryland cropping systems.http://www.ext.colostate.edu/pubs/crops/00516.html#top

Cudney, DW, Jordan, LS, Hall, AE. 1991. Effect of wild oat (*Avenafatua*) infestation on light infestation and growth rate of wheat (*Triticumaestivum*).*Weed Science* 39: 175-179.

Dahanayake, N, Ranawake, AL, Senadhipathy, DD.2014. Effects of water stress on the growth and reproduction of black gram (*Vignamungo* l.). *Tropical Agricultural Research and* Extension. 171: 45-48. DOI: http://doi.org/10.4038/tare.v17i1.5297.

DAHD 2014. *Handbook on Fisheries Statistics 2014.*Department of Animal Husbandry, Dairying and Fisheries, Ministry of Agriculture, Government of India, Krishi Bhawan, New Delhi.

Dasgupta, P, Bhattacharjee, D, Kumari, A. 2013. Socio-economic analysis of climate change impacts on foodgrain production in Indian states. *Environmental Development* 8: 5-21.

Dastagiri, MB. 2004. *Demand and supply projections for livestock products in India*. NCAP Policy Paper 21, NCAP, New Delhi

Davies, WJ, Zhang, JH. 1991. Root signals and the regulation of growth and development of plants in drying soil. *Annual Review of Plant Physiology and Plant Molecular Biology*, 42: 55-76.

Davies, WJ, Bacon, MA, Thompson, DS, Sobeigh, W, Rodriguez, LG.2000. Regulation of leaf and fruit growth in plants in drying soil: exploitation of the plant's chemical signalling system and hydraulic architecture to increase the efficiency of water use in agriculture. *J. Exp. Bot.*, 51: 1617–1626.

Davis, EF. 1928. The toxic principle of *Juglansnigra*as identified with synthetic juglone and its toxic effects on tomato and alfalfa plants. *Am. J. Bot.* 15: 620.

Derpsch, R, Friedrich, T. 2010. Global overview of conservation agriculture adoption. In Conservation Agriculture: Innovations for Improving Efficiency, Equity and Environment, PK Joshi *et al.* eds. National Academy of Agricultural Sciences, New Delhi India, p 727-744.

Dhyani, SK, Newaj, R, Sharma, AR. 2009. Agroforestry: its relation with agronomy Challenges and opportunities. *Indian Journal of Agronomy* 54: 249-266.

Dimyati, M, Mizuno, K, Kitamura, T. 1996. "An analysis of land use/ cover change using the combination of MSS Landsat and land use map: A case study in Yogyakarta, Indonesia", *International Journal of Remote Sensing*, 17(5): 931-944.

Dodd, IC. 2009. Rhizosphere manipulations to maximize 'crop per drop' during deficit irrigation. *J. Exp. Bot.* 60: 2454–2459.

Dry, PR, Loveys, BR, McCarthy, MG, Stoll, M. 2001. Strategic irrigation management in Australian vineyards. *J Int Des Sci De La. VigneEt. Du. Vin.* 35:129–139.

Du, T, Kang, S, Zhang, J, Li, F and Yan, B. 2008. Water use efficiency and fruit quality of table grape under alternate partial root-zone drip irrigation. *Agric Water Manage.* 95: 659–668.

FAO, 2014. OECD-FAO Agricultural outlook 2014. OECD, Food and Agricultural Organization (FAO), 323 p.

Farshbaf Jafari, Pirzad, S, Tajbakhsh, A M, Ghassemi, GK. 2014. Effects of Water Supply and Plant Density on Leaf Characteristics of Amaranth *AmaranthusCaudatus* L *In* 2nd International Conference on Sustainable Environment and Agriculture IPCBEE vol. 76. IACSIT Press, Singapore DOI: 10.7763/IPCBEE. 2014. V76. 4.

Ferris, R, Ellis, RH, Wheeler, TR, Hadley, P. 1998. Effect of high temperature stress at anthesis on grain yield and biomass of field grown crops of wheat. *Annals of Botany* 82: 631-639.

Fischer, RA. 1985. Number of kernels in wheat crops and the influence of solar radiation and temperature. *Journal of Agricultural Sciences* 105: 447-461.

Forest Survey of India (FSI). 1999 & 2013. India state of forest report. Ministry of Environment and Forest, Govt. of India.

Franke, AC, Singh, S, McRoberts, N, Nehra, AS, Godara, S, Malik, RK, Marshall, G. 2007. Phalaris minor seedbank studies: longevity, seedling emergence and seed production as affected by tillage regime. *Weed Research* 47:73-83.

Gathala, M K, Ladha, JK, Saharawat, YS, Kumar, V, Kumar, V, Sharma, PK. 2011. Effect of tillage and crop establishment methods on physical properties of a medium-textured soil under a seven-year rice-wheat rotation. *Soil Sci. Soc. Am. J.* 75: 1851–1862.

Glantz, MH, JH, Krenz. 1992. Human Components. In: K.E. Trenberth (editor), Climate System Modeling, University Press, Cambridge.

GOI. 2014. 19th Livestock census. All India Report. Ministry of Agriculture, Department of Animal Husbandry, Dairy and Fishery, Govt. of India, New Delhi.

GOI. 2014. Pocket Book of Agricultural Statistics, Department of Agric& Cooperation Directorate of Economics & Statistics, Govt. of India, New Delhi.

Gosain, A K, Rao, SS, Basuray, D. 2006. Climate change impact assessment on hydrology of Indian river basins. *Curr. Sci.* 903:346-353.

Gosain, AK, Rao, SS, Arora, A. 2011. Climate change impact assessment of water resources of India. *Curr. Sci.* 101(3):356-371.

Grimes, DW, Walhood, VT, Dickens, WL. 1968. Alternate-furrow irrigation for San Joaquin valley cotton. *California Agri.* 22: 4-6.

Gupta, JJ, Bujarbaruah, KM. 2005. Promising non-conventional feed and forage resources of NEH region for livestock and poultry feeding. Technical Bulletin No. 16 Published by ICAR-RC-NEH Region Meghalaya India pp.27.

Gupta, N, Kukal, SS, Bawa, SS, Dhaliwal, G S. 2009. Soil organic carbon and aggregation under poplar based agroforestry system in relation to tree age and soil type. *Agroforestry Systems* 76: 27-35.

Haileslassie, A, Blummel, M, Clement, F, Descheemaeker, K, Amede, T, Samireddypalle, A, Acharya, N S, Venkataradha, A, Ishaq, S, Samad, M, Murthy, M V R, Khan, M A. 2011. Assessment of the livestock-feed and waternexus across mixed crop-livestock systems intensification gradient: An example from the Indo-Ganga basin. *Experimental Agriculture* 47: 113-132.

Hernández, M, Echarte, L, Maggiora, A D, Cambareri, M, Barbieri, P, Cerrudo, D. 2015. Maize water use efficiency and evapotranspiration response to N supply under contrasting soil water availability, *Field Crops Research*. Vol. 178: 8-15. http://dx.doi.org/10.1016/j.fcr.2015.03.017.

Hooda, N, Gera, M, Andrasko, K, Sathaye, J, Gupta, MK *et al.* 2007. Community and farm forestry climate mitigation projects: case studies from Uttaranchal, India. *Mitigation and Adaptation Strategies for Global Change* 12: 1099-1130.

Hossain, MM, Lam, HM, Zhang, JH. 2015. Physiological Response of Soybean Genotypes to Water Limiting Conditions. *African Crop Science Journal*. 233: 227-238.

Heydari, N. 2014. Water productivity in agriculture: challenges in concepts, termsand values. Irrig. and Drain. 63: 22-28.

ICAR, 1985. Nutrient requirement of livestock and poultry. Publication and Information Division, Indian Council of Agricultural Research, New Delhi.

ICAR. 2010. Degraded and Wastelands of India, Status and Spatial Distribution. Directorate of Information and Publications of Agriculture, Indian Council of Agricultural Research, Krishi Anusandhan Bhavan I, Pusa, New Delhi 110 012.

IPCC, 2007. Climate Change 2007: The Physical Science Basis. Contribution of Working Group I to the Fourth Assessment Report of the Intergovernmental Panel on Climate Change [Solomon, S D Qin, M Manning Z. ChenM. Marquis KB. Averyt M Tignor and HLMiller. edts. Cambridge University Press Cambridge. United Kingdom and New York NY USA 996 pp.

IPCC, Intergovernmental Panel on Climate Change. 2007. Climate Change: Synthesis Report; Contributions of the Working Groups I II and III to the fourth Assessment Report of the IPCC Geneva Switzerland pp 104.

Islam, A, Ahuja, LR, Garcia, LA, Ma, L, Saseendran, A S, Trout, TJ, 2012c. Modeling the impact of climate change on irrigated corn production in the Central Great Plains. *Agric. Water Manage.* 110:94-108.

Islam, A, Ahuja, LR, Garcia, LA, Ma, L, Saseendran, AS. 2012b. Modeling the effect of elevated CO_2 and climate change on reference evapotranspiration in the Semi-arid Central Great Plains. *Trans. ASABE* 556:2135-2146.

Islam, A, Shirsath, PB, Kumar, SN, Subhash, N, Sikka, AK, Aggarwal, PK. 2014. Use of Models in Water Management and Food Security under Climate Change Scenarios in India. In: L.R. Ahuja, L. Ma, and R.J. Lascano, editors, Practical applications of agricultural system models to optimize the use of limited water. Adv. Agric. Systems Model. 5. p. 267-316. ASA-SSSA-CSSA, Madison, WI. (ISBN: 978-0-89118-344-0).

Islam, A, Sikka, AK, Saha, B, Singh, A. 2012. Streamflow response to climate change in the Brahmani River basin, India. *Water Resour. Manage.* 26: 1409-1424. doi: 10.1007_s11269-011-9965-0.

Jasti, PV, Rao, CS, Venkateswarlu, B, Singh, VP. 2014. Livelihood diversification through agroforestry in semi-arid regions of India. World Congress on Agroforestry, Delhi.

Jones, HG. 1992. Plants and microclimate: A quantitative approach to environmental plantphysiology, 2nd ed. Cambridge University Press, Cambridge UK. 428 p.

Jones, HG. 1998. Stomatal control of photosynthesis and transpiration. *Journal of Experimental Botany* 49: 387-398.

Jordan, WR, Miller, FR. 1980. Gentic variability in sorghum root systems: Implications for drought tolerance. In: Turner N C and Kramer P J. (eds) Adaption of plants to water and high temperature stress, Wiley-Interscience, New york, pp 383-399.

Kang, S, Liang, Z, Hu, W, Zhang, J. 1998. Water use efûciency of controlled alternate irrigation on root-divided maize plants. *Agric Water Manage* 38: 69–76.

Kang, SZ, Hu, XT, Du, TS, Zhang, J, Jirie, P. 2003. Transpiration coefficient and ratio of transpiration to evapotranspiration of pear tree *Pyrus communis* L. under alternative partial rootzone drying condition. *Hydrological Processes.* 17: 1165–1176.

Kaur, B, Gupta, SR, Singh, G. 2002. Carbon storage and nitrogen cycling in silvipastoral system on sodic soil Northwestern India. *Agroforestry Systems* 54: 21-29.

Kavikishore, PB, Hong, Z, Miao, GH, Hu, CAA, Verma, DPS. 1995. Over expression of P5CS increases proline production and confers osmotolerance in transgenic plants. *Plant Physiology* 108: 1387-1394.

Khan, M A. 2010. Enhancing water productivity through multiple uses of water in Indo Gangetic basin. Paper published by Innovation and Sustainable Development in Agriculture and Food Montpellier France. HAL Id: hal-00533384; https://hal.archives-ouvertes.fr/hal-00533384

Kobata, T, Okuno, T, Yamamoto, T.1996. Contributions of capacity for soil water extraction and water-use efficiency to maintenance of dry matter production in rice subjected to drought. *Japanese Journal of Crop Science*. 65: 652- 662.

Korwar, GR, Prasad, JVNS, Rao, GR, Venkatesh, G, Pratibha, G, Venkateswarlu, BJC. 2014. Agroforestry as a Strategy for Livelihood Security in the Rainfed

Areas: Experience and Expectations. Dagar *et al.* eds. Agroforestry Systems in India: Livelihood Security & Ecosystem Services, Advances in Agroforestry 10, DOI: 10.1007/978-81-322-1662-9_5, _Springer India 2014.

Kotur, SC, Satisha, GC. 2011. Zinc nutrition of horticultural crops with emphasis on yield and quality. *Indian Journal of Fertilizers* 7: 128-139.

Kotur, SC. 1998. Evaluation of lime boron and their residue on three cropping sequences of non-cruciferous vegetables for yield, composition of leaf and soil properties on an alfisol. *Indian Journal of Agricultural Sciences* 68: 718-721.

Kotur, SC. 2012. Management of acid soils for vegetable production in India. The 8th International symposium on plant-soil interactions at low pH. GKVK, Bengaluru. Pp.65-78.

Kriedmann, PE, Goodwin, I. 2003. Regulated deficit irrigation and partial rootzone drying. *Irrigation insights no. 4, Land and Water Australia,* Canberra, 102p.

Kumar, AK. 2010. Carbon sequestration: underexploued environmental benefits of Tarai agroforestry. *Indian Journal of Soil Conservation* 38: 125-131.

Kumar, BM.1997. Bamboos in the Home gardens of Kerala - a shrinking resource base. *J Non Timber For Prod,* 4: 156–159.

Kumar, R, Solankey, SS, Singh, M. 2012. Breeding for drought tolerance in vegetables. *Vegetable Science* 39(1): 1-15.

Kumar, V, Singh, S, Chhokar, RS, Malik, RK, Brainard, DC, Ladha, JK. 2013. Weed management strategies to reduce herbicide use in zero-till rice-wheat cropping systems of the Indo-Gangetic Plains. *Weed Technology* 27: 241-254.

Lal, R. 2004. Soil carbon sequestration in India. *Climatic Change.* 65 277-296.

Lal, R. 2009. The potential for soil carbon sequestration. Agriculture and climate change : An agenda for negotiation in Copenhagen. 2020 Vision Focus 16 Brief 5 IFPRI Washington DC USA.

Lal, R. 2006. Enhancing crop yields in developing countries through restoration of soil organic carbon pool in agricultural lands. *Land Degrad. Dev.* 17: 197-206.

Lee, KH. 1991. Wetlands change detection methods investigation. EPA environmental monitoring systems laboratory, Las Vegas, Nevada. pp. 42.

Leib, B, Caspari, H, Redulla, C, Andrews, P, Jabro, J. 2006. Partial root-zone drying and deficit irrigation of 'Fuji' apples in a semi-arid climate. *Irrig. Sci.* 24: 85–99.

Lemma, B, Kleja, D B, Olsson, M, Nilsson, I. 2007. Factors controlling soil organic carbon sequestration under exotic tree plantations: A case study using the CO_2 Fix model in southwestern Ethiopia. *For. Ecol. Manage.* 252: 124–131.

Levitt, J. 1980. Responses of plants to environmental stresses. Water, radiation, salt and other stresses (Vol-2). Academic Press, Newyork, pp 93-108.

Lillesand, TM, Kiefer, RW. 2001. Remote Sensing and Image Interpretation, John Wiley and Sons.

Litsinger, JA. 1979. Major insect-pests of rain-fed wetland rice in tropical Asia. *Int. Rice Res. Newsl.*, 16: 25-26.

Liu, F, Song, Zhang X, Shahnazari, A, Andersen, M N. Plauborg, F. Jacobsen, SE, Jensen, CR. 2008. Measurement and modeling of ABA signaling in potato (*SolanumtuberosumL.*) during partial root-zone drying. *Environ. and Experimental Botany*, 63: 385-391.

Liu, H, Jiang, G M, Zhuang, H Y, Wang, K J. 2008. Distribution, utilization structure and potential of biomass resources in rural China: With special references of crop residues. *Renewable and Sustainable Energy Reviews* 12: 1402-1418.

Longley, P, Donnay, J, Barnsley, M. 2001. "Remote Sensing and Urban Analysis", Taylor and Francis, London, (eds.), pp.117.

Loveys, BR, Stoll, M, Dry, PR, McCarthy, MG. 2000. Using plant physiology to improve the water use efficiency of horticultural crops. *ActaHorticulturae*, 537: 187-197.

Lucas, MR, Ehlers, JD, Huynh, BL, Diop, NN, Roberts, P, Close, TJ. 2013. Markers for breeding heat tolerant cowpea. *Molecular Breeding* 31: 521-536.

Maikhuri, RK, Semwa, RL, Rao, KS, Singh, K, Saxena, KG. 2000. Growth and ecological impacts of traditional agroforestry tree species in Central Himalaya India. *Agroforestry Systems* 48: 257-272.

Malik, RK, Singh, S. 1993. Evolving strategies for herbicides use in wheat : Resistance and integrated weed management. In : *Proc. Int. Symp. Indian Soc. Weed Sci.*, Hisar, India, 18-20 Nov. 1993. Vol. I : 225-238.

Malik, RK, Singh, S. 1995. Littleseed canary grass (*Phalaris minor Retz.*) resistance to isoproturon in India.*Weed Technol.* 9(3) : 419-425.

Malik, RK, Yadav, A, Singh, S, Malik, RS, Balyan, RS, Banga, RS, Sardana, PK, Jaipal, S, Hobbs, PR, Gill, G, Singh, S, Gupta, RK, Bellinder, R. 2002. Herbicide Resistance Management and Evolution of Zero-tillage – A Success Story. Research Bulletin, CCS Haryana Agricultural University, Hisar, India, 43pp.

Manigbas, NL, Lambio, LAF, Madrid, LB, Cardenas, CC. 2014. Germplasm innovation of heat tolerance in rice for irrigated lowland conditions in the Phillipines. *Rice Science* 21(3): 162-169.

Meinzer, FC, Goldstein, G, Grantz, D A. 1990. Carbon Isotope Discrimination in Coffee Genotypes Grown under Limited Water Supply. *Plant Physiology*. 92 1: 1 130-135. doi: http:/ / dx. doi. org/ 10. 1104/ pp. 92. 1. 130.

Mendelsohn, R. 2014. The impact of climate change on agriculture in Asia. *Journal of Integrative Agriculture*, 13: 660-665.

Michael, K, Jens, K, Annette, P, Simone, K, Katrin, D. 2015. Water productivity of poultry production: the influence of different broiler fattening systems. *Food and Energy Security* 41: 76-85.

Ministry of New and Renewable Energy Resources. 2009 www.mnre.gov.in/ relatedlinks/biomassrsources.

Mishra, JS. 1997. Critical period of weed competition and losses due to weeds in major field crops. *Farmers and Parliament* XXXIII (6): 19-20.

Mishra, RK, Pandey, VK. 1998. Intercropping of turmeric under different tree species and their planting pattern in agroforestry systems. *Range Manag. Agroforest*. 19: 199–202.

Mittal, S. 2007. Can horticulture be a success story for India.Working Paper No. 197, Indian Council of Research on International Economic Relation.

Molden, D, Murray-Rust, H, Sakthivadivel, R, Makin, I. 2003. A water-productivityframework for understanding and action. In: Kijne, JW, Barker R, Molden D(Eds.), Water Productivity in Agriculture: Limits and Opportunities for Improvement.CABI Publishing and International Water Management Institute, Wallingford,UK/Colombo, Sri Lanka

Molden, D. 1997. Accounting for Water Use and Productivity, SWIMPaper, 1-16. International Irrigation Management Institute,Colombo, Sri Lanka.

Molden, D, Oweis, TY. 2015. Pathways forincreasing agriculturalwater productivity. Water use and productivityin a river basin. IWMI Part 3 Ch. 4-7.

Morison, JL, Baker, NR, Mullineaux, PM, Davies, WJ. 2008. Improving water use in crop production. *Philosophical Transactions of the Royal Society* (London), 363: 639-658.

Matsuno, Y, Nakamura, K, Masumoto, T, Matsui, H, Kato, T, Sato, Y. 2006. Prospects for multifunctionality of paddy rice cultivation in Japan and other countries in monsoon Asia. *Paddy Water Environment* 4: 189–197.

Nair, PK, Kumar, B M, Nair, V D.2009. Agroforestry as a strategy for carbon sequestration. *J. Soil Sci. Plant Nutr*. 172: 10–23.

Nair, PKR. 1993. An introduction to Agroforestry. Department of Forestry, University of Florida, Gainesville, Florida, U.S.A., 489 p.

Narain, P. 2008. Dryland management in arid ecosystem. *Journal of the Indian Society of Soil Science* 56: 337-347.

Nath, V, Das, B, Sikka, A K, Singh, R V, Kumar, S, Rai, M and Kumar, M.2003. Fruit crop based multi-storied system for management of upland watersheds in eastern India. Abstracts and Proceedings, National Conference on Resource Conserving Technologies for Social Upliftment, 7-9 December, 2004, NARC Complex New Delhi, p. 370.

Nath, V, Das, B, Yadav, M S, Kumar, S, Sikka, A K. 2007. Guava – A Suitable Crop for Second Floor in Multi-Storied Cropping System in Upland Plateau of Eastern India. *Acta Hort*. 735: 277-95.

Nelson, D E, Repetti, PP, Adams, TR, *et al*., 2007. Plant nuclear factor Y (NF-Y) B confer drought tolerance and lead to improved corn yields on water limited acres. *Proceedings of National Academy of Sciences* 104(42): 16450-16455.

Nene, Y L. 2006. Indian pulses through the millennia. *Asian Agri History* 10: 179-202.

Newaj, R, Dhyani, SK. 2008. Agroforestry for carbon sequestration: Scope and present status. *Indian Journal of Agroforestry* 10: 1-9.

Nicholas, ME, Gleadow, RM, Dalling, M. 1985. Effect of post anthesis drought on cell division and starch accumulation in developing wheat grains. *Annals of Botany* 55: 433-444.

Nover, L, Scharf, KD, Neumann, D. 1983. Formation of cytoplasmic heat shock granules in tomato cell cultures and leaves. *Molecular Cell Biology* 3: 1648–1655.

NRCAF. 2007. Perspective Plan Vision 2025 National Research Centre for Agroforestry, Jhansi, Uttar Pradesh.

OConnell, MG, Goodwin, I. 2007. Responses of Pink Lady apple to deficit irrigation and partial root-zonedrying: physiology, growth, yield, and fruit quality. *Aus. J. Agric. Res.* 58: 1068–1076.

Oerke, EC. 2006. Crop losses to pests: Centenary review. *Journal of Agricultural Sciences* 144: 31-43.

Oki T, Kanae S. 2006. Global hydrological cycles and world water resources. *Science*, 313 (5790):1068–1072.

Pachauri, R. 2013. Conclusions of the IPCC Working Group I Fifth Assessment Report AR4 SREX and SRREN 11 November, 2013, Warsaw, Poland.

Passioura, JB. 1982. Water in the soil-plant-atmosphere continuum. In: Lange O. L., Nobel P. S., Osmond C. B. and Ziegler H. (eds) Encyclopedia of Plant Physiology, New Series Vol 12 B Springer-Verlag, Berlin, pp 5-33.

Passioura, JB. 1983. Roots and drought resistance. *Agricultural Water Management*. Vol. 7 1-3: 265-280. http://dx.doi.org/10.1016/0378-3774(83)90089-6.

Pathak, H, Bhatia, A, Jain, N. 2010. Inventory of greenhouse gas emission from agriculture. Report submitted to Ministry of Environment and Forests, Govt. of India.

Pathak, MD. 1968. Ecology of common insect pests of rice. *Annl Rev Entomol*. 13: 257-294.

Pattanayak, A. 2014. Challenges in crop improvement under climate change with special reference to water deficit and high temperature. In: Saikia U.S., Ramesh T., Ramkrushna G. I., Krishnappa R., Rajkhowa D. J., Venkatesh A. and Ngachan S. V. (eds) Natural Resource Management for Enhancement of Adaptation and Mitigation Potential under Changing Climate, ICAR Research Complex for NEH Region, Umiam, Meghalaya, India, pp7-14.

Pattanayak, SK, Misra, BK. 2012. Management of acid soils for sustainable rice production in India. The 8th International symposium on plant-soil interactions at low pH. GKVK, Bengaluru. Pp.45-64.

Pattanayak, SK, Rao, DLN, Mishra, KN. 2007. Effect of biofertilizers on yield, nutrient uptake and nitrogen economy of rice-peanut cropping sequence. *Journal of the Indian Society of Soil Science* 55: 184-189.

Peden, D Freeman, A, Astatke, A, Notenbaert, A. 2006. *Investment options for integrated water-livestock-crop production in sub-Saharan Africa*. ILRI research paper #1. Nairobi: ILRI.

Peden D, Taddesse G, Misra AK. 2006b. *Water, livestock and human development*. IFPRI, IUCN, IWMI. 2005. "Let it Reign: The New Water Paradigm for Global Food Security." Final Report to CSD-13. Stockholm International Water Institute, Stockholm.

Pinheiro, HA, DaMatta, FBM, Chaves, ARM, Loureiro, ME, Ducatti, C. 2005. Drought Tolerance is Associated with Rooting Depth and Stomatal Control of Water Use in Clones of Coffeacanephora. *Annals of Botany*. 96: 101-108.

Prajapati, N D, Purohit, S S, Sharma, A K and Kumar, T. 2003. A Handbook of Medicinal Plants. Agribios India, 553 p.

Prasad, PVV, Boote, KJ, Allen, LHJR, Sheehy, JE, Thomas, JMG. 2006. Species, ecotype and cultivar differences in spikelet fertility and harvest index of rice in response to high temperature stress. *Field Crops Research* 95: 398–411.

Prasad, R, Prasad, D. 2010. Evaluation of bio- efficacy of newer insecticides against major lepidopterous insect pests of rice ecosystem at Rachi. *J.Appl. Zool. Res.*, 21(1): 25-27.

Preap, V, Zalucki, MP, Nesbitt, HJ, Jahn, GC. 2001.Effect of fertilizer, pesticide treatment, and plant variety on realized fecundity and survival rates of *Nilaparvatalugens*Stål Generating Outbreaks in Cambodia". *J. Asia Pac. Entomol.*, 4 (1): 75–84.

Price, AH, Cairns, JE, Horton, P, Jones, HG and Griffiths, H. 2002. Linking drought-resistance mechanisms to drought avoidance in upland rice using a QTL approach: progress and new opportunities to integrate stomatal and mesophyll responses. *J. Exp. Bot.* 53: 989–1004.

Priya, A, Nema, A K, Islam, A.2014. Effect of climate change and elevated CO_2 on reference evapotranspiration in Varanasi, India - A case study. *J. Agrometeorology* 161: 44-51.

Qiming Zhou. 1995.The Integration of GIS and Remote Sensing for Land Resource and Environment Management. Proceedings of United Nation ESCAP Workshop on Remote Sensing and GIS for Land and Marine Resources and Environment Management, School of Geography, The University of New South Wales, Sydney, NSW 2052, Australia.Feb 13-17, 1995, pp.43-55.

Rai, P, Chaturvedi, P O, Singh, R, Singh, U P. 2009. Biomass production in multipurpose tree species in natural grasslands under semi arid conditions. *Journal of Tropical forestry* 25: 11-16.

Raturi, GB, Chadha, K L. 1993 Horti-Silvi-Pastural system. In: Advances in horticulture Vol.2, Fruit Crops. Edt. KL Chadha and OP Pareek. Malhotra Publishing House New Delhi pp. 805-11.

Reichardt, W, Brioes, A, de Jesus, R, Padre, B. 2001. Microbial population shifts in experimental rice systems. *Applied soil ecology*: 17: 151-163.

Reichardt, W, Mascarina, G, Padre, B, Doll, J. 1997. Microbial communities of continuously cropped irrigated rice fields. *Applied and Environmental Microbiology*. 63 (1): 233-238.

Reissig WH, Heinrichs EA, Litsinger JA, Moody K, Fiedler L, Mew TW, Barrion A T. 1986. Illustrated Guide to Integrated Pest Management in Rice in Tropical Asia. International Rice Research Institute, Philippines, 411p

Rietveld, WJ. 1983. Allelopathic effects of juglone on germination and growth of several herbaceous and woody species. *J. Chem. Ecol.* 9: 295–308.

Rosenberg, NJ, Mckenney, MS, Martin, P. 1989. Evapotranspiration in a greenhouse-warmed world: a review and a simulation. *Agric. For. Meteorol.* 47: 303-320.

Sadras, VO. 2009. Does partial root-zone drying improve irrigation water productivity in the field? A meta-analysis. *Irrig. Sci.* 27: 183–190.

Saha, R, Tomar, JMS, Ghosh, PK .2007. Evaluation and selection of multipurpose tree for improving soil hydrophysical behaviour under hilly eco-system of north east India. *Agroforestry Systems* 69: 239-247.

Saied, SM, Saied, JA. 2001.The effect of rice and tomato residues on the growth and some of the yield components of some wheat cultivars *Triticum aestivum* L. *Al-RafedainSci.J.* 13: 31-44.

Saini, HS, Aspinal, D. 1982. Abnormal sporogenesis in wheat (Triticumaestivum L.) induced by short periods of high temperature. *Annals of Botany* 49: 835-846.

Saini, HS, Sedgley, M, Aspinal D. 1984. Developmental anatomy in wheat of male sterility induced by heat stress, water defcit or abscisic acid. *Australian Journal of Plant Physiology* 11: 243-253.

Samadi A, Sepaskhah AR. 1984. Effects of alternate furrow irrigation on yield and water use efficiency of dry beans. *Iran Agric. Research*, 3: 95-115.

Samra JS, Singh SC. 2000. Silvopasture systems for soil, water and nutrient conservation on degraded land of Shiwalik foothills sub-tropical northern India. *Indian Journal of Soil Conservation* 28: 35-42.

Sanyal, SK, Bhattacharya, K. 2012. Chemistry of acid soils of India-Distribution, properties and management for sustainable crop production. The 8th International symposium on plant-soil interactions at low pH. GKVK, Bengaluru. Pp.15-44.

Schon, MK, Einhellig FA. 1982. Allelopathic effects of cultivated sunflower on grain sorghum. *Bot. Gaz.* 143(4): 505-10.

Seckler, D. 1996. The New Era of Water Resources Management: From 'Dry' to 'Wet' Water Savings. Research Report 1. International Irrigation Management Institute(IIMI), Colombo, Sri Lanka.

Seckler, DD, Molden, R,Sakthivadivel. 2003. "The Concept of Efficiency in Water Resources Management andPolicy." In J.W. Kijne, R. Barker, and D. Molden, eds., Water Productivity in Agriculture: Limits and Opportunities forImprovement. Wallingford, UK, and Colombo: CABI Publishing and International Water Management Institute.

Sepaskhah, AR, Sichani, SA. 1976. Evaluation of subsurface irrigation spacings for bean production. *Canadian Agricultural Engineering*, 18: 23-26.

Serraj, R, Sinclair, TR. 2002. Osmolyte accumulation: can it really help increase crop yield under drought conditions? *Plant, Cell and Environment*. 25: 333–341.

Shahi, HN. 1978. Competitive effects of *Chenopodium album* for soil moisture and nutrients in wheat. Indian Pest control 20 (4),14-16Wilson, R.G. 1988.

Biology of weeds in the soil. Pp. 25-39 in M.A. Altieri and M. Liebman (Eds.), *Weed Management in Agro-ecosystems: Ecological Approaches*. CRC Press, Inc., Boca Raton, FL.

Sharma, AP, Katiha, KP. 2011.Recent advances, issues and strategies for research in inland fisheries - Perspectives of CIFRI. In Souvenir of Ninth Indian Fisheries Forum. 25-31.

Sharma, R, Sharma, E, Purohit, AN. 1995. Dry matter production and nutrient cycling in agroforestry systems of mandarin grown in association with Albiziaanf mixed tree species. *Agroforestry Systems* 29: 165-179.

Sharma, SK, Vashishtha, BB. 1985. Evaluation of jujube buffel grass horti-pastoral system under arid environment. *Ann. Arid Zone*. 24: 303-09.

Shiklomanov, IA. 1999. World Water Resources and their Use. Database on CD Rom. Paris, UNESCO.

Shiklomanov, I. 2000. International Water Resources Association Water International. 25: 1.

Shinozaki, YK, Kasuga, M, Liu, Q, Nakashima, K, Sakuma, Y, Abe, H, Shinwari, ZK, Seki, M, Shinozaki, K. 2002. Biological mechanisms of drought stress response. *JIRCAS Working Report* 23: 1-8.

Shinozaki, YK, Shinozaki, K. 2005. Organization of cis-acting regulatory elements in osmotic- and cold-stress-responsive promoters. *Trends in Plant Sciences* 10: 88-94.

Sikka, AK, Haris, AA, Upadhaya, A, Thakur, A, Reddy, AR, Dey, A, Subash, N. 2008. Water productivity assessment in the canal and tubewell commands of Bihar: A case study. *Research Bulletin No. 1*, CPWF-IGB Coordinating Unit, ICAR-RCER, Patna.

Sikka, AK, Haris, AA, Upadhaya, A, Thakur, A, Reddy, AR, Dey, A, Subash, N. 2008. Water productivity assessment in the canal and tubewell commands of Bihar: A case study. Research Bulletin No. 1, CPWF-IGB Coordinating Unit, ICAR-RCER, Patna.

Singh, HP. 1999. Integrated Watershed Management in Drylands –A System Perspective. In: Singh H P Ramakrishna Y S Sharma K L Venkateswarlu B eds Fifty Years of Dryland Agriculture Research in India. Central Research Institute for Dryland Agriculture, India, pp 259–270.

Singh, K, Chauhan, HS, Rajput, DK, Singh, DV. 1989. Report of a 60 month study on fitter production, changes in soil chemical properties and productivity under Poplar P.deltoides and Eucalyptus E. hybrid interplanted with aromatic grasses. *Agroforestry Systems* 9: 37-45.

Singh, K, Sharma, US, Swaminathan, R, Dashora, PK. 2009. Management of insect pests of *Vignaradiata* L Wilczek. *Appl. Ecol. Environ. Res*, 7: 99- 109.

Singh, OP, Sharma, A, Singh, R, Shah, T. 2004. Virtual water trade in dairy economy irrigation water productivity in Gujrat. Economic and Political Weekly July 31st, 2004.

Singh, RP. 1988. Dryland Agriculture Research in India. Pages 136-164 in 40 years of Agricultural Research and Education in India, New Delhi India. ICAR.

Singh, TP. 2003. Potential of Farm Forestry in Carbon Sequestration. *Indian Forester* 129: 839-843.

Spreer, W, Nagle, M, Neidhart, S, Carle, R, Ongprasert, S and Muller, J. 2007. Effect of Regulated deficit irrigation and partial root-zone drying on the quality of mango fruits (*Mangifera indica* L. cv. Chok Anan). *Agric. Water Manage.* 88: 173–180.

Stoll, M, Loveys, B, Dry, P. 2000. Hormonal changes induced by partial root-zone drying of irrigated grapevine. *Jour. of Experimental Botany*, 51: 1627-1634.

Stone, P. 2001. The effects of heat stress on cereal yield and quality. In: Basra A. S. (ed) Crop Responses and Adaptations to Temperature Stress, Food Products Press, Binghamton, NY, USA, pp. 243–291.

Subbarao, GV, Johansen, C, Slinkard, AE, Rao, RCN, Saxena, NP, Chauhan, YS. 1995. Strategies for improving drought resistance in grain legumes. *Critical Reviews in Plant Sciences* 14: 469–523.

Swamy, SL, Puri, S, Singh, AK. 2003. Growth biomass carbon storage and nutrient distribution in GmelinaarboreaRoxb. stands on red lateritic soils in Central India. *Bioresource Technology* 90: 109-126.

Taiz L, Zeiger E. 2006. Plant Physiology 4th edition. Sinauer Associates, Inc., Publishers, 764p.

Torres, A, Oliva, RM, Castellano, D, Cross, P. 1996. First World Congress on Allelopathy. A Science of the Future pp. 278. SAI (University of Cadiz). Spain, Cadiz.

Turner, NC. 1986. Crop water deficits, a decade of progress. *Advances in Agronomy.* 39:1-51.

Venkateswaralu, J. 2010. Rainfed agriculture in India: Research and development scenario. Indian council of Agricultural Research New Delhi: 508.

Venkateswarlu, J. 1986. Efficient Resource Management for Drylands. In: 15 years of Dryland Agriculture Research, CRIDA, Hyderabad, India, pp 42–58.

Viets, FG Jr. 1962. Fertilizers and the efficient use of water. *Advances in Agronomy* 14: 223-264.

Vikram, P, Kumar, A, Singh, A, Singh, NK. 2012. Rice: genomics-assisted breeding for drought tolerance. In: Tuteja, N., Gill, S. S., Tiburico, A. F., Tuteja, R. (eds.) Improving Crop Resistance to Abiotic Stress, Wiley- CH Verlag GmbH & Co. K Ga A, Germany, pp.713–729.

Vision 2050, 2013. ICAR Research Complex for Eastern Region.

Vorosmarty, CJ, McIntyre, P, Gessner, MO, Dudgeon, D, Prusevich, A, Green, P, Glidden, S, Bunn, SE, Sullivan, CA,Liermann, CR. 2010. Global threats to human water security and river biodiversity, *Nature*, (467):555–561.

Vyas, S and Nein, S. 1999. Effect of shade on the growth of *Cassia ungustifolia*. *Indian Forester*. 125: 407–10.

Wahid, A, Gelani, S, Ashraf, M, Foolad MR. 2007. Heat tolerance in plants: an overview. *Environmental and Experimental Botany* 61(3): 199–223.

Walker, BH, Steffen, WL. 1999. The nature of global change. In: B.H. Walker et al. editors, The Territorial Biosphere and Global Change, University Press, Cambridge, U.K.

Xiong, Y, Wang, RG, Mao, G, Koczan, JM. 2006. Identification of drought tolerance determinants by genetic analysis of root response to drought stress and abscisic acid. *Plant Physiology,* 142: 1065-1074.

Zhang, H, Oweis, T. 1999. Water-yield relations and optimal irrigation scheduling of wheat in the Mediterranean region. *Agricultural Water Management* 38: 195-211.

Zhang, ZY, Broeckling, CD, Blancaflor, EB, Sledge, MK, Sumner, LW, Wang, ZY. 2005. Overexpression of WXP1, a putative *Medicagotruncalata* AP2 domain containing transcription factor gene, increases cuticular wax accumulation and enhances drought tolerance in transgenic alfaalfa (*Medicago sativa*). *Plant Journal* 42(5): 689- 707.

ICRISAT. 2007. Pro-Poor Biofuels Outlook for Asia and Africa: ICRISAT's Perspective. (Hyderabad).

Clancy, JS. 2008. Are biofuels pro-poor? Assessing the evidence. *The European Journal of Development Research* 20: 416.

UNESDA. 2007. Small-Scale Production and Use of Liquid Biofuels in Sub-SaharanAfrica: Perspectives for Sustainable Development. Background Paper 2. United Nations Department of Economic and Social Affairs Energy and Transport Branch. Division for Sustainable Development, New York.

❑❑❑

